The Exceptional Trainer

A No Nonsense Guide for the Trainers of Emergency Communications

by
Sue Pivetta

Professional Pride
Sumner, WA 98390

THE EXCEPTIONAL TRAINER

Professional Pride
1812 Pease, Sumner, WA 98390
1-800-830-8228

This book is available by contacting Professional Pride.

ISBN 1-882960-11-4

TABLE OF CONTENTS

Where to Find Those Handy Forms

Straight Shooting

To avoid criticism,
do nothing,
say nothing,
be nothing.

Elbert Hubbard

STRAIGHT SHOOTING

Sue WHO?

These 'About the Author' things are always written by the author. Seems kinda dishonest, but then, nobody else knows that much about me—or cares. They'd have to ask me anyway. Since I don't like talking about myself as if I'm dead, I'll just tell you how I came to write this book for trainers—if that's OK.

Love Love Relationship

"I can do that," I said confidently as I cut out the help wanted ad titled Emergency Dispatcher. I had been working on the school playground for four years and it was great. I got to be the biggest kid on the field *and* I had a bullhorn! But, I wanted to do something more challenging. (Boy, I never do anything small.)

Anyway, for some reason the nasty-tempered Chief hired me despite my perky nature. (Back then I *was* perky.) I became the graveyard dispatcher of a small police department and I LOVED IT. We handled fire and medical calls, watched the jail, shuffled paperwork, and coordinated various practical jokes and 'get evens.'

Within one year we were merged into a Com Center. I was so excited! I found out my guts were bigger than my brain when I sat down to dispatch for three cities I had never even driven through. What a fantastic time that was! In four years I applied for Supervisor. What can I say, I was young, innocent, and thought I could save the world. I spent most of my time trying to save my behind!

Wild Woman

For ten incredible years I was a part of this big dysfunctional family. I went through a divorce. (What a surprise, huh?) The divorce put me in a place where I needed to evaluate my goals and my spirituality. I became this wild woman, obsessed with wanting to know stuff. *Any stuff!* Stuff I didn't know, stuff I couldn't understand. Stuff about my alcoholic parents, my abusive marriage, my dysfunctional workplace. I don't know what happened. I don't remember being hit by lightning—but it was like that. At work I became the person who did 'stuff.' I put together the first SOP. It wasn't good, but it WAS. I

A teacher affects eternity; [s]he can never tell where [the] influence stops.

Henry B. Adams

worked on the training manual, I sat on oral boards, I went to APCO meetings. I was into this profession *one hundred percent!*

Then one day, I had this great idea. We were having problems with trainee turnover. They just weren't making it through the training. Also a beautiful person (and great dispatcher) lost it, like going into quicksand. I couldn't do anything to pull her out. I began to think about everything. I began to wonder, "If we took training and moved it out of the Com room, would we have more success?" I took my idea to the director and he kicked me out of his office.

What a Trip!

Not being one to take rejection lightly, I began to whine. Someone suggested to me that I begin a college training course for dispatchers. *"No,"* I said. *"It's never been done." "Since when did that ever stop you,"* he said. So, with that challenge I contacted the local college and gave them a proposal. It took. I became college instructor.

For ten years I worked with college students of all shapes, sizes, colors, ages, and abilities. I also listened. *Who was an exceptional teacher; and who was mediocre?* I don't do mediocre, so I began to study those teachers who really changed lives, the ones that students came back to visit year after year. It didn't take much observation before I found their magic secret: they cared! They cared about teaching, about people, about having a great time.

Out There!

I was writing as I taught. Finally my Dean told me to write a book and stay away from the copier! So, I wrote my first book. Having minus-zero patience waiting for publishers to look at my material, I self published. I began Professional Pride to sell the book and worked at it part time. Gradually I created more products for the training program.

In 1994 I wrote *The 911 Puzzle: Putting the Pieces Together* for NENA. I taught this course at national conferences around the United States and Canada. This was fun!

I went back to school (as a student this time): a college called Antioch; full of intellectuals and people who are aware of the global effects on stuff. Some of the information here comes from my college learning. My focus is on adult learning, the arts, and the human condition. My BA says 'Communication,' but it should say 'Mind Expanding.'

If there is anything the nonconformist hates worse than a conformist it's another nonconformist who doesn't conform to the prevailing standards of nonconformity.

Bill Vaughan

In 1995 I quit teaching when I realized that my 'war stories' were eleven years old. I decided to do Professional Pride full time. I had the opportunity to speak to many trainers. What I was hearing? *"Our trainers need help!"* While consulting, I saw over and over that agencies weren't paying attention to the human side of training: motivation, needs, communication, evaluation, and recognition.

So here we are, writing *The Exceptional Trainer.* I say *we* because I had a lot of help from my editor, Betty Allen, an organized, articulate, and wise woman of humor.

What's Up Now?

My future goals include a distance college degree for tele-communicators, a leadership book, 'workshops-in-a-box,' and providing opportunities for workshop leaders through Professional Pride. I want the message to be of clarity, purpose, motivation, and love.

Please feel free to share your thoughts and dreams with me.

Sue Pivetta

The willingness to create a new vision is a statement of your belief in your potential.

David McNally

It's Up to You

This is a 'collection' of thoughts, ideas, tips, and discussions about training in the adult environment of an Emergency Communications Agency. This message comes from someone who has been there (and a few other places), gathered together wisdom from many sources, and put it on paper for you.

You may not agree with what I have to say in this book. That's OK. A book is just something someone is saying. Just because it's in print, doesn't mean it's absolute. These are my views based on my experiences working in this wonderful profession, training adults for twenty years, taking college courses, pursuing life-long self-directed learning, and doing research. I think if you listen you will learn. It gets a little loosey-goosey at times, but, oh well, we should have fun while we're here, don't you think?

Live, Love,

Have Fun!

and Learn!

The chief function of your body is to carry your brain around.

Thomas Alva Edison

TRAINING TERMS

(Important! Read these!)

Attitude: A combination of concepts, emotions, information, and perceptions with a judgment that results in an emotion that influences a person's behavior.

Behavior: The actions or conduct of a person which are motivated by needs or beliefs.

Feeling: To be aware of, to experience, an emotion. Physiological and/or mental reaction to an event or stimulus.

Intuition: The direct knowing of something without the conscious use of reasoning. Reaction to *feelings*, not thoughts.

Judgment: To form an idea, opinion, estimate, or decision by comparing thoughts or information.

Learning: A change in perception, behavior, or attitude brought about by acquiring new information or experience or insight.

Motivation: Those processes that can (a) arouse and instigate behavior, (b) give direction or purpose to behavior, (c) continue to allow behavior to persist, and (d) lead to choosing or preferring a particular behavior.

Need: Condition experienced in a person that leads a person to move in the direction of a goal. Some needs are stronger than others so a person acts on the strongest need.

Perception: Immediate mental grasp of situations (objects). A personal understanding based on *past* experiences or information.

Skill: The ability to do something without conscious thought about each step. To achieve success regularly. To produce the desired results repeatedly.

Looking at the Old with New Eyes

The wise leader, learning how things happen, acts accordingly. The average leader also learns how things happen, but vacillates, now acting accordingly and then forgetting. The worst leaders learn how things happen and dismiss the single principle as total nonsense. How else could their work be so futile?

The Tao of Leadership

LOOKING AT THE OLD WITH NEW EYES

> *If you see in any given situation only what everybody else can see, you can be said to be so much a representation of your culture that you are a victim of it.*
> *...Hayakawa*

Whither 'Tao' Goest

Tao Te Ching is the Chinese book of wisdom from Lao Tzu. You may have heard "*The journey of a 1000 miles begins with a first step.*" Lao said that. Tao is not a religion. One definition I found said Taoism is more interested in "*...intuitive wisdom, rather than in rational knowledge.*" Tao relies on nature to provide examples of wisdom, such as change. Transformation and change are essential in nature, yet we receive the naturalness of change with hysterical face rubbing.

Tao offers common sense. In our profession we pay great tribute to the idol of common sense. "If yer ain't got it, yer cain't get it." And we all know what happens to new hires who demonstrate any lack of common sense—pfft, out the door! Yet, you gods and goddesses of common sense, common sense is rare. I know, I am one of you.

You may have heard, or maybe read, *The Tao of Pooh*. How about the *Te of Piglet?* No? Possibly the *Tao of Physics?* Great books. The titles are attention getters for sure, because we certainly can't be expected to look for wisdom in conventional places, like the Bible, the Koran, or the Noble Truths of Buddhism.

What we like is for someone to study wisdom and common sense, buy a commercial TV spot, and then tell us what we already know but don't do anything about. Someone with inch-long white hair and leotards can motivate us when the mirror cannot.

So, why am I even introducing Tao? I think sometimes we have to go out of our comfort zone, our usual way of thinking, to grow. If you look for creativity and problem solving in everything—you will create and solve. Everywhere I looked there was Tao. I think about my profession and what can be done to fill needs. So I draw from Tao

Insanity: Continuing to do the same thing and expecting different results.

Anonymous

and everywhere else for new thought—wisdom for our collective neurosis.

This is not my wisdom, but this is my solution, my research, and my answer to finding the hidden rewards in this incredible profession. This is my way of healing, of searching, of listening, and of growing.

> *In a perfect world, not a single thing received any injury, and no living being came to a premature end. Men might be possessed of the faculty of knowledge, but they had no occasion for its use. At this time, there was no action on the part of anyone—but a constant manifestation of spontaneity. ...Taoism*

Many agencies and college programs are doing a great job of training. They didn't get there by accident—it was through self-evaluation, research, trying, failing, trying again, success, improvement. Many of these concepts you've heard or know. Some are new. Many times it's frustrating to know what is right, but feel powerless to affect real change.

And of course, it's not a perfect world, and we really need all the help we can get to become motivated to do what we know is right, to become quieted so we can hear the truth, to become tired so we give up fighting. We need help to understand and accept each other and to understand and accept ourselves.

The Brain

The supervisory center of the NERVOUS SYSTEM in all vertebrates. The brain controls both conscious behavior (e.g., walking and thinking) and most involuntary behavior (e.g., heartbeat and breathing). In higher animals, it is also the site of emotions, memory, self-awareness, and thought. It functions by receiving information via nerve cells (neurons) from every part of the body, evaluating the data, and then sending directives to muscles and glands or simply storing the information. Information, in the form of electrochemical signals, moves through complex brain circuits, which are networks of the billions of nerve cells in the nervous system. *A single neuron may* receive *information from as many as 1,000 other neurons.*

The forebrain, composed of the limbic system and cerebral cortex, regulates higher functions. The limbic system (including the thalamus, hypothalamus, pituitary, amygdala hippocampus, and olfactory cortex) is associated with vivid emotions, memory, sexuality, and smell. The forebrain's cerebral cortex, in the uppermost portion of the skull, has some areas concerned with muscle control and the senses and others concerned with language and anticipation of action.

The cerebral cortex is split into two hemispheres, each controlling the side of the body opposite to it. In addition, the right hemisphere is associated with perception of melody, nonverbal visual patterns, holistic views, events. It is diffuse in nature, non-verbal and spontaneous, playful and artistic, with no time orientation. The left hemisphere is associated with verbal skills, is analytical, linear, sequential, concrete, active, and goal oriented. It is ordered and efficient with a time orientation.

The Concise Columbia Encyclopedia

The Brain and Training Connection

Two college courses greatly influenced my thinking on our training programs. One was about learning concepts called 'No Fault, No Blame.' This three day course was taught by a former K-12 teacher named Claude Beemish. Dissatisfied with traditional school methods, he researched, wrote, and became an expert in new thinking about how humans learn. His message is clear : we are unjustly blaming learners for not learning. Just as we blame children for failing in our school systems, as trainers we blame either ourselves or the trainees for their inability to grasp what we throw at them. Much of the course was devoted to the brain and how it works. We were then taught how to use this knowledge to better facilitate the learner's brain processes.

The second course I took was 'The Brain: how it works, how we think, why we think.' This course lifted the fog from my own understanding about how we can get to a higher level of knowledge, skills, and attitudes in our profession. This course required reading *Emotional Intelligence* by Daniel Goleman, a text that further convinced me we need to change our minds about what we think (or better yet, don't think) about our training programs.

Of course I can't put the entire body of knowledge in print for you, but I can begin your questioning process. I am going to ask questions that will begin your own higher level thinking about what you are doing in your training.

> *True learning comes from inquiry, the key to all real learning. Learners have to ask questions to define the key issues or topics of their learning. They have to ask questions about such issues to explore all the possibilities and ramifications. The inquiry process means they can touch base with several aspects of a topic, covering its scope at first broadly and then more particularly. By asking questions, the learner chooses the most appropriate, meaningful response to resolve, if possible, a relevant aspect of learning.*
>
> *...Norma Goldstein,* Using Inquiry to Promote Learning

With these brief facts, I hope you will first understand that the body of information I have put forth for you on the care and feeding of a trainee has evolved from my own introduction into the complexity of the human mind and human emotions. Learning involves entire

For years we blamed inadequate 'hiring' for the turnover problem, when in reality a great deal of blame goes to the training program.

Pivetta

processes in the mind and is promoted or blocked by emotions. If you wish for a prototype for learning about mind and emotions, look no further than yourself. How do *you* learn?

- The first amazing realization I had was—the brain does much more than 'think.' Consider the following business taking place in our brain when it is asked a question, fed information, asked to perform a new task, or to think a new thought.

    We analyze, break into parts, probe, examine, process, compare, contrast, perceive, interpret, challenge, distinguish, discover, note, store, reject, accept, explain, define.

    This started me thinking: How many ways can we challenge or provoke the learner to use all processes available in our training programs to develop better thinkers?

- I watched a video where the brain is 'thinking' and color could be seen with special detectors. It showed that different 'types' of thinking involved different levels of activity in distinct areas of the brain.

    The highest level in remembering and processing—expressing and decisions—happens in the higher levels of the brain called the neo-cortex.

    Which made me wonder: What type of thinking are we provoking with our ways of training in the Com Center?

- I learned the brain is 'triune' which means we have several layers of the brain: the neo-cortex which does higher levels of thinking; the limbic system which handles body, emotions, play, memory, and the day-to-day thinking; and the basal ganglia or reptilian brain which handles survival, anger, sex—the baser emotions.

    We move to different parts of this triune brain at times by downshifting. Downshifting sends us from the higher levels of processing, analyzing, etc., to the lower levels of fighting, survival. Definitely not *the learning area!*

    This made sense to me! We downshift our learners in many ways. This made me ask: How we can keep from downshifting the learner?

- Next I put together information from Maslow's Hierarchy of Needs with downshifting from Claude Beemish and related it to

Abraham Maslow (màz' lo)
1908-1970
American psychologist and a founder of humanistic psychology who developed a hierarchical model of human motivation, in which a higher need, ultimately that for self-actualization, is expressed only after lower needs are fulfilled.

The American Heritage Dictionary of the English Language

the many times when learners experienced a block, especially in agency training.

Abraham Maslow stated that we cannot move from lower level areas of need to higher levels of learning unless conditions are right. For example, if you are worried about where to live, you can't pay much attention in school. Each level offers greater security and acceptance, and allows the person to attend to higher level thinking—like that done in the neo-cortex.

OK, so I'm wondering: If we pull our trainees into the lower level areas of worry, insecurity, and safety, are they thinking about survival instead of processing?

- I learned that everything we have ever heard, seen, thought, or felt is in our brain—somewhere. I learned when the brain is touched in a certain place by a scalpel, it evokes a particular memory. But when that brain part is removed, the memory stays.

This leads to another book called The Holographic Universe and the Holographic Brain *by I. Don Noh offering information about emotions and memory as holographic in nature—that is, they move around in the electricity of the brain.*

Now I am wondering: If everything is up there, what can we do to create Telecommunicators who can not only absorb the information, but put all the pieces together when needed to come up with good judgment and fast action?

- I found out there is something called Cognitive Perception Reality (CPR for short). This means that what we *believe* we see is what we see. We apparently look for reinforcement for our held beliefs and we have some difficulty 'changing our minds.' What we believe comes from programming—either what we did to ourselves, or what our parents or environment did to us.

Seeing is not believing. Believing is seeing. A national survey found that during the Presidential debates people did not change their minds about the candidates. They were looking for evidence that their candidate was right!

Creative minds always have been known to survive any kind of bad training.

Anna Freud

[AMEN!]

I am wondering: What are our trainees believing, without our permission. What do they hear, interpret, and then begin to believe in the training process?

Conclusion

> *...the more brain mind, body emotion connections we pay attention to, the more we expand our consciousness and increase the flexibility and fluency of mental functioning. ...Bob Samples,* Open Mind, Whole Mind.

And the more our trainers believe that the complexity of learning should be considered in the design and implementation of training, the more higher level of critical thinking 911 Telecommunicators we will turn out.

We are going to look at the training program, the trainers, adult learning, creative training ideas, problem areas, and evaluations. Included in the information in this book is the continuing spark of thought that we must use new eyes in the care and feeding of our trainee's process of learning.

Does Your Training Program Measure Up?

If you keep doing what you're doing, you'll keep getting what you're getting.

Pivetta

DOES YOUR TRAINING PROGRAM MEASURE UP?

The Premise

Exceptional training programs pull the learners in by allowing them to participate in designing and directing their own learning: involving them, and creating an atmosphere of responsibility and acceptance. Who knows better what they know and don't know? I am always better at putting together some disassembled toy if I am given a picture of what it's supposed to look like at the end—we all are. It is the ideal program that allows for practice, feedback, needs assessment, and goal setting. *And* the ideal program has trainers who are empowered to be creative, dynamic, and responsible for design and implementation.

Considerations in Program Development

Is there a perfect model of on-the-job training? Maybe not perfect, but what about effective or functional? Each agency is different, so each has its own resources, needs, and goals. However, (I love saying 'however') there are identifiable components that turn a rigid program into an attentive program—attentive to the needs and the best use of resources. This model is adapted from Training Magazine, *How to Design the Ideal Training Course*, by David D. Cram. I find it very thought provoking (and my thoughts love to be provoked). So, I added 'points' because they came out of my pointy head. Material in italics comes from the article.

12 Features of an Ideal Training Program

1. *An ideal course will make it clear to everybody just what the end product skill will be.*

   ➻Point #1:

   > Remember, the trainee is watching a veteran. They need to know what is expected of *them*, and by when. They already understand they won't have the speed, expertise, or flow of an experienced dispatcher—but HOW skilled

Humans are the only animal that can think how they think.

Whole Brain Thinking
Wonder and Donovan

must they be by WHEN. Short term objectives are the answer to providing a map, a goal chart for the trainee.

➣Point #2:

Read again and notice the word EVERYBODY. Clear to *everybody*—that means the trainer and the entire agency. Training shouldn't be like making fudge, "OK, I *think* you're ready."

2. *An ideal course will let everybody know how the objective is going to be tested.*

➣Point #1:

Testing may be stressful, but not knowing or not being able to show your stuff is worse. Testing may seem like a dirty word, but most of us don't know any other way to find out if someone knows something. There are other ways: feedback forms, communication, observation, outside observers, etc.

➣Point #2:

Many people have the knowledge or skill, but are test phobic. Using small evaluations leading up to the same final is fair and effective. This way you know your trainee isn't failing testing—they are failing the skills.

3. *An ideal course will take advantage of what you know and can do when you begin.*

➣Point #1:

I'm so glad he said this 'cause I have a caution! Laterals may or may not know what you think they know. Each person you hire will have a different set of skills coming in. If Mary worked at an ambulance company she may or may not have the knowledge you require—always evaluate first.

➣Point #2:

It is worth the time and effort to get to know your trainee before you begin training. Give the new hire a questionnaire of your own making.

4. An *ideal course will give the student as many choices of paths through the instruction as possible.*

The future does not get better by hope, it gets better by plan. And to plan for the future we need goals.

Jim Rohn

➻Point #1:

You are allowed to be creative with learning on console training! For example, if someone is really having a hard time with map reading, send them out riding, or ask them what they think would help.

➻Pointy #2:

If a trainee tells you they would benefit from a certain type of assistance (for example, they want to ride), *believe* it. Most adults have a handle on their own learning process.

5. *An ideal course will provide a range of training aids.*

➻Point #1:

Note taking, observation, feedback, tapes, videos, reading, writing, study, projects. There are times when a trainee and trainer need some time apart. This time is ideal for self-directed learning.

6. *An ideal course will allow enough time for any qualified individual to finish.*

➻Point #1:

The power word here is 'qualified.'

➻Point #2:

Many times training programs are cut short because the trainee is 'needed.' Nothing could be worse from a liability standpoint. As for the trainee, who has missed a portion of the training, this shortened training is a very uncomfortable place to be.

7. *An ideal course will provide a map to keep everyone informed about where the program is in the system.*

➻Point #1:

Although it may seem like a lot of work to lay out a program week by week, it is worth the initial development time. What a great feeling to know you are on track—or even ahead of it! If you and your trainee are behind, this is a great time to make some changes, ask for additional time, or problem solve.

Whatever does not spring from a man's free choice, or is only the result of instruction and guidance, does not enter into his very being, but still remains alien to his true nature; he does not perform it with truly human energies, but merely with mechanical exactness.

Karl Wilhelm Von Humboldt

↣Point #2:

Make sure to provide your administrator with progress reports on where you are with your trainee.

8. *An ideal course will provide opportunity for practice of the skills being taught.*

↣Point #1:

I am the champion of training simulators in the Com Center!

9. *An ideal course will provide feedback to the student on a regular basis.*

↣Point #1:

Oooh, I love it when I say something and then an expert says the same thing! (See the chapter on evaluations.)

10. *An ideal course will test often and in non-threatening ways.*

↣Point #1:

That nasty word 'test' again. Feedback knowledge, evaluate skills, ask for reassurance that what you thought was absorbed was absorbed, find gaps. Gaps are natural.

11. *An ideal course accepts feedback from the student for self correction.*

↣Point #1:

OOOH, again, I'm telling you, it's important! Use*: I saw, I felt, I learned, I still need to know.* It's wonderful.

12. *An ideal course compares student to objective, not student to student.*

↣Point #2:

This is the hard one. You will have students who learn quickly and then come to a learning curve and slow down. You may have students who learn slowly but steadily; and others that freeze. Just allow for differences in learning. It's a moving, living, breathing thing, and no two experiences are alike.

It's always good to know—a lobster does not share his food. And, of course, it's because he's shellfish.

Pivetta

Conclusion

The material on the ideal program was probably written for college courses. Still, these adult learning concepts can easily be applied to agency training.

To take an honest look at your program, complete the two agency training analyses on the following pages. Then proceed to our suggestions on how to shape better training.

TRAINING PROGRAM DYSFUNCTIONAL RATING

Rate the following on the scale: 1 = Never Happens 5 = Consistently Happens

1	2	3	4	5	Your trainer wasn't thrilled about being asked to train–or refused.
1	2	3	4	5	Trainees were not sure what they would be learning and when.
1	2	3	4	5	Training was interrupted because of other priorities.
1	2	3	4	5	The procedures were not consistent with: training, what others said, what the trainee saw others do.
1	2	3	4	5	The training was unorganized
1	2	3	4	5	Trainer had many other responsibilities.
1	2	3	4	5	Trainer wasn't sure about correct current or updated procedures.
1	2	3	4	5	Little or no written material or related reading was offered.
1	2	3	4	5	Evaluations were late or non-existent.
1	2	3	4	5	Trainee had more than three trainers.
1	2	3	4	5	Training was cut short.
1	2	3	4	5	Lost a good candidate.
1	2	3	4	5	Hired a bad candidate.
1	2	3	4	5	Didn't release a poor candidate.

Lowest score: 14; Highest score: 70.
The lower the score the more effective the program.

AGENCY TRAINING CONSISTENCY ANALYSIS

Rate the following on the scale: 1 = Never Happens 5 = Consistently Happens

1	2	3	4	5	All employees in the same position perform general tasks of the job in the same way.
1	2	3	4	5	Every supervisor/trainer will give the same explanation of how each task should be done.
1	2	3	4	5	It takes employees approximately the same amount of time to perform a specific task.
1	2	3	4	5	All employees use the same process to perform identical tasks.
1	2	3	4	5	All employees follow well-defined procedures for the agency.
1	2	3	4	5	Field personnel compliment communications about consistency.
1	2	3	4	5	Similar quality standards are consistently attained by all employees and supervisors..
1	2	3	4	5	All employees are evaluated by the same consistent standards.
1	2	3	4	5	The definition of a good performance is understood by all members and is used as a basis for performance appraisals.
1	2	3	4	5	Supervisors are consistently and uniformly rated according to high standards.

Lowest score: 10; Highest score: 50.
The higher the score the more consistent.

Shaping an Ideal Program

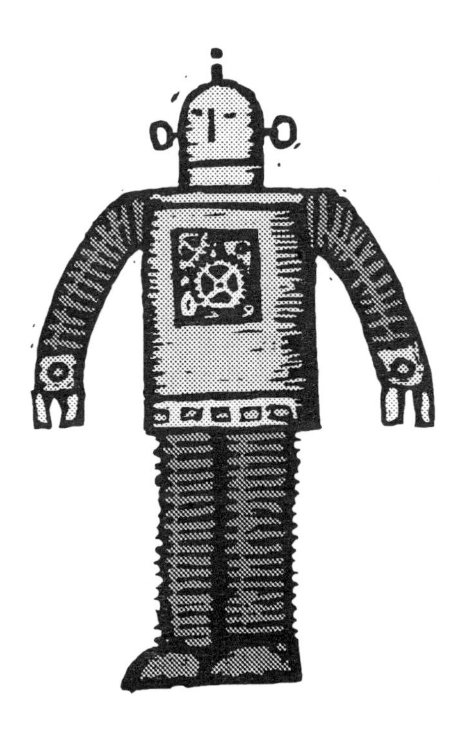

Our attitude towards ourselves should be 'to be satiable in learning' and towards others 'to be tireless in teaching.'

Mao Zedong

SHAPING AN IDEAL TRAINING PROGRAM

Let's start basic. Your agency may have a great training program, or your agency may have no organized training program, or your agency may be somewhere in between—some organized and some disorganized training, with you in the middle. Wherever you are, there is a workable and simple process to improve your agency training program.

Four Steps to Improvement

Step One:	Create the Training Syllabus
Step Two:	Construct an Evaluation Process
Step Three:	Develop Trainers
Step Four:	Allow Trainers Time to Prepare

Step One: Create the Training Syllabus

How was your program created? Your agency may or may not have used a formal analysis. If they didn't use a formal process, they certainly had to use an informal process. There is no way around it. You have to determine what it is you need to train before you can begin a training program. The only difference between formal and informal is, of course, that formal is thorough and organized. Formal also prepares a foundation for written guidelines and every other organized, documented process. Formal is an occupational analysis.

First, complete an occupational analysis. (See form) This sounds lengthy and cumbersome, it was for those of us who first did this. You, however, have the advantage of using ours with very little adjustment.

An Occupational Analysis is a method of breaking down any job type into training units. This procedure is done by college courses to determine what training units are needed to turn out a skilled worker. Ask the following questions:

1. What do these people do?

NOTE: The State of Oregon hired a researcher and sent questionnaires to each dispatch agency in the state. Each was asked what duties and tasks were done and how often. With this information they were able to determine what they needed to present in their certification training.

2. How important is this task or how often do they do this?
3. What are the steps to learning this task?

These questions are answered with a DACUM, or Developing a Curriculum, process. (A curriculum is what you will be teaching.) The DACUM process uses experts from the field in structured brainstorming sessions with a facilitator so an absolute training program can be developed. This also should be done in an agency.

How an occupational analysis works:

1. Break the job down into major components; for example, with us it could be phones, radios, computers, and records.
2. Next, break down each major component into categories. Phones will become emergency phones, non-emergency phones, ring down lines, cellular calls, and whatever other types of lines you may attend.
3. Then take a category (such as emergency phones) and break this down into 'tasks.' Emergency phones tasks could include: answering domestic violence calls, assaults, and so on, until each emergency phone call type is accounted for.

After this process, you will have a complete list for the job. Note: If you have different job classifications, and therefore different training for Call Taker and Dispatcher, you may need to do an analysis for each classification.

This process proceeds from the very general to the very specific. It's surprising how many different aspects of the job are discovered with this method. It's a great way to write a job description or justify that pay raise!

After you complete your Occupational Analysis—Aha! You now have an outline for your training units or objectives.

Each Communications entity is different, so an occupational analysis is customized. Your agency may include more records functions, front counter work, prisoner search, monitoring cameras, security doors, or a variety of other duties.

Write objectives. Once you have the occupational analysis, you have the skeleton for your objectives. For each 'task' you will write an objective. An objective will state what you will accomplish with this unit of training.

We are drowning in information, but starved for knowledge.

Haislitt

A training objective is simply a goal. What is it we want to learn about domestic violence calls? Teachers write objectives for their own sanity, so it is clear what they want to achieve. A teacher begins each day by looking at a particular objective, sharing the objective, and ends by again looking at the objective using a measure to determine if it was accomplished. If not, OK, it can be reached the next day.

Your objective doesn't have to follow academic models. The format should be kept simple so it will be used. No matter how you accomplish it, writing some style of objective is called *Directed Console Training.*

Providing Directed Console Training. Your console training can be as organized as any classroom. Although you can't always count on a particular event to happen (so you can train by example), you can cover any topic completely by being prepared. Using prep sheets accomplishes organization and documentation, and reduces the chance a subject will be overlooked. This method creates accountability. With organized, directed training, the information is potent, meaningful, and documented.

Develop lesson plans. From your objectives you can now complete a lesson plan. This is for the trainer, to guide them on what to teach and how.

Prepare units of instruction. A unit of instruction is several lesson plans given a timeline and some type of measure of learning.

This 'measure of learning' does NOT mean a test or evaluation. I do not believe a test is the best feedback you can receive on what a person has learned. To me a 'test' or 'quiz' or whatever you want to call it, is simply cramming information down someone's throat and expecting them to regurgitate it. (I picture these baby pigeons we used to feed Vet's dog food!) Instead, I want to know what they keep, what they understand, where the gaps are, what they still need to know. Anyway, we will discuss this many times throughout the book.

List types of learning experiences. Use your creativity, knowledge, experience, and the learner's knowledge of their own learning style to create more than one way to learn. For example: you may want to teach about pursuits. You could do the age old, *"Tell them what you want them to know and expect them to get it,"* or you could do something that makes a lasting impression on the learner. For example, give them a tape of a pursuit and the procedure for pursuits. Ask them to evaluate the tape according to the procedure. Next practice pursuits. Have them listen to and complete the Professional

We don't need more strength or more ability or greater opportunity, we need to use what we have.

Kellogg

Pride In Pursuit tape. The learning experience must be interactive. They share learning, you share your experience, knowledge, and insight.

Gather training resources such as books and tapes. Learning can be self induced. A learner can be involved actively in their own learning by reading, research, and evaluation. Gather together many resources. Direct the learner on what you expect them to get out of the material. For example, ask the learner to find everything related to E911 in an APCO Bulletin magazine and write an article on what they read. You may want to use some of these learning exercises when you must be absent.

Identify outside training or trainers. Don't forget to use the experts in this field. You may send your student to spend some time with the Fire Training Officer, or the person in charge of training records at the police department. You may want to call in a local college instructor to put on a small class, inviting other centers or some of your experienced people. You can also use the hospitals; for example, a tour of the burn center.

Write the training manual. You may already have a training manual (that would be great). The training manual works in coordination with the SOP. For example, on the pursuit. You will have an SOP on pursuit—this says 'Thou Shalt....' In other words, the rules. A training manual takes the SOP one step further and says, 'Thou Shalt..., and here's how you do that.' Writing a training manual is a lengthy process, but once it's done, it's done (with only minor changes according to changes in SOP). Within your training manual you should have knowledge checksheets, daily feedback forms, examples, and resources.

OK. Step One was to create a training syllabus. By now you should have all the components of a training program: objectives, lesson plans, training manual, forms, resources, and creative ideas for providing a high level learning experience for your trainee.

Step Two: Construct an Evaluation Process

One basic truth about adult motivation: people need to know how they are doing. In my college training course, a student would want to meet with me. *"How am I doing?"* The cute little face so somber. This was an indication to me I wasn't keeping up with feedback and evaluations. If one person were asking, there were probably others who didn't have the guts to ask. Again, adult learners and employees

It's what you learn after you know it all that counts.

Wooden

want feedback from their trainers or employers. The evaluation process is vital.

Evaluations are a management tool necessary to maintain a high level of efficiency and safety. Can an evaluation fall short of filling these significant needs? They can and often do. Ask yourself if you look forward to your next evaluation. Ask a co-worker how 'valuable' their last evaluation was. Does your evaluation process actually assess? Is your evaluation form a learning tool? Do your employees come out of an evaluation process energized and motivated? Is the process geared toward ongoing quality improvement for your agency? (See Chapter 11, 'Do You Love Evaluations?')

Step Three: Develop Trainers

Compensation

"We, who have done so much for so long for so little." This could be funny if it weren't so true. Our trainers are not compensated properly. I am still asked to train for nothing, and *I* do this for a living! Like it or not, compensation is a measure of worth. It is in every other part of our society, so why not here? Trainers should be compensated for their work. A Director told me once he was upset because his trainers wouldn't attend a 911 training conference that was in town. He was even willing to let them use their comp time! I knew he had once been a Police Captain so I asked him, *"If these were Police Trainers would they have to take their own time, or would you pay them to go?"* We both knew the answer. Again, trainers should be compensated for their work *and* they should be paid to attend training. Period. (So, Sue, tell us what you *really* think!)

Hiring Trainers

I believe trainers should be hired, not appointed; *unless* your department is small and it is known up front that training will be part of the job duties. If a person is to be hired and it is known they will eventually be training, take that into consideration and include the ability to become a trainer in the hiring process. Can *anyone* be a trainer? Not if they don't want to, not if they don't have patience, love for people, knowledge, a process, a system, and understand adult learning. Get what I am saying? *Please* don't throw people into being trainers.

'Paradigm' is a new buzzword. It means a model or example to follow—a new way of ideal thinking.

Pivetta

Training Trainers

I believe many times we get desperate and send our people to just any old training because we don't have the wherewithal to check it out first. Ask for references, follow up. In my workshops, I ask students to write a letter to their boss thanking them for the training and giving their view of the class.

Develop a Train the Trainer program in house. If you are a small department, get together some magazine articles, books, tapes, real life accounts from previous trainers. Hook up the new trainers with old trainers or outside trainers. In other words, put some effort into providing the resources inside—empower your own department.

Initiate a Support/Follow Up System

Don't leave me now! This is the real step into a new paradigm! Meetings, budget, support. The care and feeding of trainers—a neglected thing 'til now. Trainers must be able to continue their education, and seek out support (SOS) for their immediate and long term problems. Allow them to create a vision and allow this vision to become a reality. Fewer trainers will burn out if there is a system of support.

Step Four: Allow Trainers Time to Prepare

Time seems to be the most valuable tool for the trainer—and the hardest to get. Trainers not only need time to organize and prepare, but to present, communicate, evaluate, and counsel. Ask trainers to develop a schedule where they have a minimum of one hour a day off the console. If you must pay OT—so be it!

Conclusion

The Four Steps to Training Program Improvement don't require any substantial financial investment. The four steps do involve time—and time is our most precious and guarded commodity. The initial improvement takes a great deal of time. I'm asking for this. Time for on-going attention to constant improvement is also necessary.

Next, some real help for the new trainer.

Occupational Analysis

Sample Outline

✷ **Duties: Major job requirements**

Answer Phones
Operate Computer
Operate an Emergency Radio
Provide Reports

✷ **Clusters or GAC, General Area of Competence**

(Duties are broken down in manageable number of tasks.)

For example: *Answer Phones* - what kind of phones?

✷ Emergency Phones
Non Emergency Phones
Ring Down lines
Direct alarm phones
Administrative lines

Emergency Phones - what kind of emergency phones?

Medical emergency
✷ Police emergency
Fire emergency
Crisis emergency

Police emergency phones - what kind ?

Bank Alarm
Domestic Violence
Outside agency assist

and so on...

Sample Lesson Plan

Unit Title: Non Emergency Phones

Unit Topic: Animal Control Calls

Items to be Covered:

1. Nuisance Animal Complaints
2. Dangerous Animal Complaints
3. Dead or Injured Animals
4. Livestock
5. Game

Performance Objectives:

Given the information on the role and responsibility of Animal Control and Law Enforcement responses to animal complaints, the trainee will correctly answer questions by the trainer regarding what action should be taken on the five types of animal calls.

Student Learning Activities:

1. Read SOP on Animal Calls
2. Discuss information with trainer
3. Tape review of animal complaint problem
4. Review of animal control log book
5. Visit Animal Control when riding

Student Evaluation:

Written (oral) exam identifying proper responses to animal calls.

Resource Material Needed:

SOP/ Animal control guidelines / Animal Control Book
Tape / 15 minutes for evaluation

Time Allocated:

Two hours

Comments:

Make sure both #s to animal control are covered.

FILLING OUT THE DIRECTED TRAINING SHEET

I. Prepare and Motivate:

A basic teaching concept. The trainer discusses the objective for the day and why it is important. This may be obvious to you, but the trainee wants a sense of how often this happens and what risk there is in failing to accomplish this task properly. You can use war stories, horror stories, examples, question and answer sessions, or articles to make your point.

II. Background Knowledge:

The trainer relates the topic to whatever prior knowledge, training, skill, or information has been presented.

> ***Note**: Does your agency provide classroom training? If so, have you seen your trainees' evaluations? No classroom training? You're IT, huh. What does your trainee know?. Review their job application, talk to them about their knowledge, give a test.*

III. Procedures, Policies and Methods:

Review the written procedure for this call type. Explain policy regarding this call type for radio and phones. Cover common methods of successfully completing this task according to policy. Procedures don't tell you how you are going to handle every possible problem.

IV. All Possible Situations:

Discuss every deviation that could occur with this situation. Play "What If...." This allows the trainee a chance to cover a lot of ground.

Example: Pursuits All Possible Situations
Accidents / Poor Radio / Losing Contact

V. Assessment of Knowledge:

Identify that the trainee has been provided with all information regarding this call type. Document what was discussed and attach a copy of the map or notes with signatures. You may want to provide a quiz or assignment.

VI. Positive Model:

The trainee should have a chance to see an individual effectively doing this task. You can't simply call up a pursuit for practice (unless you have a simulator). You can have a collection of pursuit tapes and require the trainee to leave the room and review the tapes, chart the pursuit, and give you feedback on the procedure followed. Call it *active listening.*

VII. Skill Practice:

How do you know you have skill? When you don't have to think much about it and have a great deal of success at it. Remember the first time you typed? You had to think about it and you had a great many errors. With time it became comfortable, your speed increased, and you made fewer errors. You must have a simulator for your trainees to practice call receiving, CAD, and radio. It isn't much trouble to get it together. The first pursuit they take should always be in a simulation setting.

Sample Directed Training Sheet

Date: ____

Trainer/Trainee: ________________________________

Subject: ______________________________________

I. Prepare and Motivate:

____

II. Background Knowledge:

____

____

III. Procedures, Policy and Method:

____

____

____

____

____

IV. All Possible Situations:

____

____

____

____

V. Assessment of Knowledge Provided Y/N

____

VI. Positive Model Provided Y/N

____

VII. Skill Practice Provided Y/N

____

Additional Comments:

____

____

____

____

Trainer Initials _____ Trainee Initials _____

Yikes! I'm a Trainer. Now What?

The first and great commandment is:
Don't let them scare you.

Elmer Davis

YIKES! I'M A TRAINER. NOW WHAT?

A good educator is someone who makes hard things easier.

Who Is this Exceptional Trainer?

Insensitivity or unfamiliarity with empathy in training is one of the culprits in a chaotic training experience. It is quite possible many trainers do not understand that adult learning is about *emotion.* Learning isn't about gaining information—it's about how you think and feel about the information you're getting.

There's a lot to learn about training: how to write an objective, how to organize training. What is lacking is not how to put together training, but how to hold together training. To bring our training to a higher level of critical thinking we must dig deeper into the human component of training and then use all that other stuff.

An Exceptional Trainer knows:

- how people learn, blocks to learning
- how to develop and evaluate skills
- how to evaluate knowledge and attitudes
- how a trainer can affect a learner's attitude
- how to recognize needs
- how and why motivation is important to the adult learner

Above all else, exceptional trainers know how to communicate their own thoughts, feelings, needs, and perceptions. Exceptional trainers *listen* for the thoughts, feelings, needs, and perceptions of the trainee and their peers. An exceptional trainer notices and acts.

YOU Can Be Exceptional!

I am large, I contain multitudes.

Walt Whitman

Steps for Trainers Once the Program Is in Place

I've asked you to be an exceptional trainer, but the truth is all this talk about feelings and needs isn't concrete enough for the new trainer. The new trainer is usually an exceptional Dispatcher who one day is given the heave-ho into some old growth forest without a guide or any provisions. What can be done? What real steps can be taken to get you, the new trainer, started? The following are steps in the right direction—toward excellence!

Step One:	Develop a Schedule
Step Two:	Introduce Training Program to Trainee
Step Three:	Address Trainee's Concerns
Step Four:	Prepare and Motivate

Step One: Develop a Schedule

If you don't provide your learners with a map, how will they know where they are going, the expected time it will take to get there, what their rest stops are, and how will they know they've arrived?

Develop a training schedule—but be flexible! You have a set amount of time, and within that time the exceptional program will allow a student to accelerate or extend. Learning is not a static process—it is a moving, changing, living dynamic that has many components.

The learners will want to know if they are on course. Set out landmarks, goals, and deadlines; but make it clear they can get there early or late—it's a goal, not a brick wall. Flexibility reduces tension and anxiety, and creates a better atmosphere for learning.

There will be times when the schedule will be a mess. You know this is true, so why not surrender and create some alternative plans for every possible scenario. The trainer will need to work around vacations, doctor appointments, illness, overtime, and even their own work schedule.

Plan a schedule as you would a map. In addition to having a main route, plan alternate paths as well.

Step Two: Introduce Training Program to Trainee

This is sitting down with the trainee and going over the entire process. Include your expectations; request trainee feedback on their feelings, desires, fears, and ideas; and provide clarification as needed.

"We have a great training program," rookie said. "Every day my trainer gives me a DOR [Daily Observation Report] about what I learned."

"Oh," I said, "And what do you give your trainer back in the way of what you observed or learned?"

Pivetta

The trainee should be given the following: a Trainer Bio and/or Yearbook, rules and regulations with comments, present schedules for training, and guidelines for evaluations.

Get to know the person. Let them know something about you and those they will be working with. Ask them if there is anything they need. Ask them how they learn best. Ask them what they already know. Ask them if they are worried about anything. ***Listen***.

Make your expectations clear. Allow the trainee the opportunity to understand the results of their actions. If the trainee is tardy, what will happen? If the trainee is not progressing, what will happen? etc. etc.

Introduce yourself to the trainee. Give them your experience, your thoughts on training, your hopes, and your excitement for their learning. And it's not there, they will feel it.

Step Three: Address Trainee's Concerns

Open and extend a communications avenue. How does the person tell you something is wrong, how they are feeling, what they are learning? Just sitting down and talking is good, but not documented. I am going to propose one of the most valuable training tools ever designed. (Probably the most helpful suggestion in this book!)

This is the **Daily Feedback Form.** (See example) Learners will move to a higher level of processing information by filling out a feedback form. They may fill out this form after shift, at home that night—it doesn't matter, but it must be turned in to the trainer before the next training session. The trainer then takes the form, writes their own comments, and uses it as a discussion tool for learning. This is now *documentation* of what the trainee is seeing, feeling, learning, and needing. What a deal! No more miscommunications.

Another great training tool is a **portfolio**. Require the trainee to keep a portfolio of their learning: feedback forms, evaluations, tests, quizzes, riding forms—everything will be handed in at graduation (or probation, or whenever). This is a very valuable book. Allow the trainee to add their own essays on how their learning progressed. When you review their portfolio, you will have a clear image of their learning experience.

Step Four: Prepare and Motivate

You must sell training to your learner every day. I know you may think this is unnecessary, but think about your own learning. Don't you question training: What will I be learning? Will it be relevant? We

Learning does not only occur when the trainer says so, it occurs all the time—in the Com Room, in the lounge, during the ride-along. So, I ask, "What is your trainee learning without your permission?

Pivetta

often have blocks to learning. Preparing and motivating removes the first block: Why do I need this?

Sell your entire program and each lesson as vital and 'learnable.' I say this because often trainers feel the need to scare the pants off the trainee with horror stories about how many people failed in the last go-around. (Now there's a real motivational leader!) An adult learner must believe they can learn, not that it's an uphill climb they may not make. What if your hiking guide told you, *"Four out of five people who take this hike don't make it up the hill."* You might say, *"I'll be the one."* But that's you, it may not be your trainee. What you want to say is, *"This is do-able, you are smart, I'm a great trainer, this is going to be challenging and fun, and most of all, I BELIEVE IN YOU!"*

We've been talking about the learner and some of the things you can do to prepare. Now, what is the newest thinking about the adult learning process?

Daily Feedback Form

DATE:

What I saw …	What I thought/felt …
What I learned …	What I still need to know …

Trainee ____________________ T/O ____________________

Confidential Learning Assessment

Trainee: ______________________________________

Trainer:____

Dates: ____________________ to ____________________

1. What do you feel are your highest achievements here?

2. What do you feel are your best attributes on the job?

3. What was the easiest skill for you to master in this work?

4. In what area(s) of knowledge do you feel the most comfortable?

5. What makes you a good employee?

6. Are you aware of any improvement needed in your skills?

7. Are you comfortable with your level of knowledge.

   Explain:

8. What could we do to assist you to be your best?

9. What areas could we improve upon in training?

10. What is the most difficult part of this job?

11. What would you change here if you could?

12. What is the best part of working here?

13. List any errors you have made in your work here?

14. When do you learn best here?

15. What hampers your learning here?

SAMPLE ASSIGNMENT SHEET

Trainee: ____
Trainer: ____
Date Assigned: ______________________________________
Due Date: ____

Objective:

Example: Become familiar with TTY calls

Assignment:

Student will watch TTY video and make 3 practice TTY calls.

Tasks to be competed:

1. Watch 20 minute training video
2. Complete TTY worksheet provided
3. Using trainer TTY make three practice calls

Evaluation Tool:

TTY worksheet completed
Printout of three TTY calls

Bringing Adult Learning Theory Home

Caring about the trainee's needs is not about babying the trainee. It's about getting the most of your training dollars! Allowing the learner to be in a better mental state for learning.

Pivetta

Bringing Adult Learning Theory Home

When an adult takes on the invitation to learn something, the experience can be a challenging, invigorating, or even profound event for both the learner and the facilitator. Since the early eighties there have been many theories on adult learning needs and styles that claim a trainer can elevate the success of the learning adventure by understanding the trainee and designing the course for success. We can also do that in agency training.

The following pages are a collection of articles on adult learning and this profession. Each examines a current body of knowledge and relates it to this profession in a way that provokes thinking. This is just one method of getting discourse going about concepts outside our normal realm of thoughts. If you keep doing what you're doing, you'll keep getting what you're getting. I am always looking for improvement, not necessarily because something is wrong, but because change is growth. To improve training shows a willingness to enrich your thoughts, expand your skills, and grow.

Learning Styles

Learning styles is an interesting subject. There is no doubt that how a person learns affects their feelings about a particular course or class. Are styles that important in the learning process? From my own observation of adult trainees in a vocational education setting, from my involvement with on-the-job training at Communications, and from my work as a professional workshop leader, I believe there are learning needs that are constants and that ensure a learning experience regardless of a person's *style.*

Adults expect all learning experiences to be 'relevant to the job.'

Pivetta

7 VITAL COMPONENTS TO ADULT LEARNING

In my research, I have found that there are seven components which are vital to adult learning. You can easily apply these to your own agency training program.

1. *Adult learners need or can use the information.*
2. *Adult learners must respect and receive respect from others during the learning process.*
3. *The adult learning process must be a collaboration.*
4. *Adult learners learn best when they can relate the material to their own thoughts and experiences.*
5. *Adult learners must be ready to learn and have a comfortable learning environment.*
6. *Adult learners cannot be forced to learn.*
7. *The adult learner's motivation to learn is related to their thoughts about their ability to succeed.*

1. *Adult learners need or can use the information.*

The learner must feel that the learning is personally useful to them. For this to happen the materials and information must be at an appropriate level. Many times we assume that people have knowledge which they may or may not have.

Adults as selective learners

Adult learners seek education for a variety of reasons: work, pleasure, volunteering, personal growth. Although the adult trainee may attend a class voluntarily, but if they discover the information will not be useful and directly apply to their needs, they may mentally or physically pull out of the session. Because adults view their time as valuable, most people will not tolerate waste. Adults are generally cautious about choosing classes to attend and take time to carefully read a course description or talk to the instructor to ensure a fit. According to the National Center for Research in Vocational Education, working with adults means accepting that the learner has many other time consuming roles (e.g., spouse, parent, union

If learning is a 'change' in behavior, how do you know when learning has occurred? By observation, communication, listening, and noticing.

Pivetta

member, scout leader, athlete, worker, business owner). Training that is perceived as not relevant will cause the trainee's mind to wander.

Adults want to be autonomous about what they learn, how it is learned, and when and where they learn. With so many choices on how to spend their valuable time, adults need enough information to determine the value of any potential training. One example of this on console training is the failure of the trainer to provide good learning experiences during times when the trainer is not present. Another is to have the trainee observe without some idea of what it is they are looking for. Trainees want to use training time constructively.

Mandatory attendance and learning

How does mandatory or required training affect adult motivation? Although some adults may not be motivated to put energy and effort into mandatory training, there may be a long-range incentive besides just learning. Adults are goal oriented. Although attending mandatory classes may be seen as forced learning, often completing a mandatory class to reach a particular goal (such as a certificate) is enough motivation for the adult to buy into the training. Simply getting the certificate or pleasing the employer may be enough to solicit genuine attention from the learner. Trainees' goals are their own.

As a facilitator of industry workshops, I sometimes encounter an employee who was 'sent' to learn and feels their time is being wasted. They sit arms crossed, stubbornly refusing to take part in or enjoy the class. In an effort to motivate this learner, I spend the first thirty minutes playing tapes and offering personal examples of why this learning could be beneficial. I try to engage the learners with questions, soliciting their input on their expectations and their learning needs. I relate to them what I want to accomplish and why I feel it is worthwhile for them to spend their time with me. Since I began this practice, I have seen an improvement in the cohesiveness of the groups I facilitate.

Prepare and motivate

Although the learners come to the training hoping they can use the information, they anticipate the trainer's validation when the session begins. Relevance is important for the learner. It is the task of the trainer to relate the material to the trainee's learning needs. This basic teaching premise is called 'prepare and motivate.' *"Why do I need this stuff?"* is the learner's first thought when preparing to receive learning. The objective of the course, related by the trainer, must meet with the adult's perceived needs.

I can't memorize the words by themselves.
I have to memorize the feelings.

Marilyn Monroe

I understand that you may feel 'keeping the job' should be enough motivation. It is on a large scale. But, day to day, lesson to lesson, you want to connect the learner to the lesson so there are no gaps if the trainee unknowingly buys out.

What should I get from reading this?
What do I gain by riding?
What am I watching for while observing?

Are you selling your lessons?

2. Adult learners must respect and receive respect from others during the learning process

The adult learner must feel respect in the learning environment both for and from the trainer and other trainees, as well as those in the Com Center.

Respect for the learner

Study of the adult learner indicates that respect is a principal component in achievement. The trainer sets the social environment for the trainee. The slightest contempt felt by the trainer toward a trainee can be perceived not only by the trainee, but by others. If the trainer has negative feelings toward the trainee, both the trainee and the class will recognize it. The trainer may also lose support from others by appearing intolerant.

I can recall many trainees who didn't fit in because of their appearance or personality. My interaction with these people was closely watched by the group. At times my acceptance would help the trainees fit in, but other times nothing could help.

Adults are often internally motivated by feelings of worth, self esteem, and achievement. When an adult senses judgment, non-acceptance, discrimination, and/or rejection from the trainer or the other trainees, the ability to learn may be obstructed. In the Com Center, trainees are concerned about fitting in and are very sensitive to the feelings of the group toward them. As a trainer are you noticing this need?

Respect for ideas and attitudes

The learner must feel safe in a challenge to personal bias, behaviors, or ideas. Speaking out should not lead to personal denigration. If this

How about this: We give a trainee as much support as possible and <u>then</u> see if they 'survive.'

Pivetta

happens, a learner will mentally challenge information but not openly express their questions or differences, thus nullifying the learning opportunity. Learning takes place by comparing new information to old. If there is no opportunity to clarify, the chance to understand is missed. At times the trainee may seem to challenge the material and this is not acceptable to the trainer. The trainee is seen as difficult, or a know-it-all. Help the trainee to formulate questions that are less confrontational. Do you have the confidence to allow challenge?

A safe atmosphere

In workshops that have a mixture of Telecommunicators and managers, there is often an atmosphere of censure and retaliation. Questions result in silence and discomfort because people do not want to express their real opinions. According to M. S. Knowles, *"Fear of communicating leads to a valueless atmosphere for imparting information with no critical evaluation."* The ability to risk without loss is vital to the adult learner's active involvement in the class.

Another aspect of the safe atmosphere is to feel safe to admit ignorance. To some adult learners, admitting they don't know something is shameful. This comes from previous environments where the trainee was ridiculed or when a trainee feels a sense of low self esteem in the group setting. There are two ways this trainee may appear to the class: either silent, or argumentative and defensive. The silent person appears not to have much to contribute. The argumentative, defensive learner seems like a know-it-all who incessantly maintains their expertise instead of listening. Do you encourage questions or ideas without judgment?

Respect for the trainer

In an academic setting, trainees judge the ability of the trainer. *"She's great!" "He's boring."* Respect is not only for the trainer's subject knowledge, or skill at delivering instruction, but also the trainer's attitudes toward learning and the learner. R. L. Wlodkowski describes this combination of competence and caring as the *four essential characteristics of a respected trainer:* expertise, empathy, enthusiasm, and clarity. Do you have a balance of competence and caring?

Vision is a process that allows you to think ahead to where you want to be and what you want to be doing, and to create workable plans to lead you there.

Fred Pryor

3. *The adult learning process must be a collaboration*

> *Although parties have different roles at different times, the learner must believe both roles are equal in value and both work toward clear objectives.*

The trainer as facilitator

Adult learners invest time, money, and personal sacrifice when pursuing an education. They expect the trainer or teacher to assume the role of enthusiastic guide through the process. The trainer must not only understand the subject and the learning goal, but must be willing to offer feedback. Adults want to know how they are doing so they can do their best. (You've heard this before in this material, and you will hear it again!)

Although adults appreciate the opportunity to be autonomous in their learning environment, they need their efforts to be appreciated. If adults are given a learning task, they want recognition for their work, either from the group or from the trainer. An example is when the instructor gives the group an exercise, but doesn't allow time to share the results. Adults need to feel their contributions are valuable. Do you return to the group?

Collaboration in the group

The trainee also sees group collaboration as meaningful. Adults want to hear other's thoughts and have the opportunity to offer their own insights. A trainer can encourage a constructive group discussion by using active lecture and dialogue techniques. To feel included in the group as an active participant is vital to the 'critical thinking' learning that group experience is meant to effect. Do you encourage exchange of thoughts?

Collaboration in learning

On a larger scale, the trainee's ability to effect a needed change in the training program brings collaboration full circle. An avenue of communication between the learners and the trainers can bring about a sense of community and growth. Trainees who feel their needs, ideas, thoughts, and vision have value will feel a sense of community and self worth. The feeling that we are all equal with different jobs but the same goal can add to the excitement of the training program. This is going to be a difficult concept for agencies to swallow! It says you should ask trainees how to improve your training program. Do you ask the learner for suggestions?

Silences regulate the flow of listening and talking. They are to conversations what zeros are to math; crucial nothings without which communication can't work.

Goodman

4. Adult learners learn best when they can relate the material to their own thoughts and experiences

> *To reach each trainee's art of learning, the process must involve a variety of activities that explore higher level thinking by allowing reflection, analysis, and evaluation.*

Time to connect and process

Listening is not learning, telling is not teaching. By comparing, connecting, contrasting, paralleling, examining, exploring, and scrutinizing the information offered, the learner will shape the learning into a meaningful experience. Whether it's during on-the-job training, in college classrooms, or during workshops and seminars, trainees don't *own* the information unless they are given the time and permission to work with the idea, concept, data, or material. The educator should use a variety of learning experiences, both group and individual. This doesn't suggest that traditional classroom techniques are not valuable, but they can be overdone when used as the only way to impart information.

Lecture has often received bad press in the classroom, but there is value in expert led discussion. An effective lecture is called an 'active lecture,' a type of learning that offers a valuable interchange. One example of non-learning in lecture is when the trainer asks a question then immediately provides the answer before the listeners have a chance to formulate a response. Soon the trainees expect that the trainer will provide the information and take no initiative to mentally explore the question. They are released to wander in their thoughts.

> *Higher order thinking tasks such as analysis, synthesis, and evaluation involve strategies that involve the learner with the material through active learning and thinking about what they are doing and where the learning is going. ...Bonwell and Eison,* Active Learning

I had a young woman come into my college class wondering if this training was right for her. She asked if I lectured. I told her there was lecture every morning after break. She was so relieved. She was dropping out of her first choice of training because she felt the teacher, who she believed had a great deal of expertise in her subject, did not share her knowledge with the class.

There is no more noble occupation in the world than to assist another human being—to help someone succeed.

Alan Loy McGinnis

Learning styles and relating

S. F. Brookfield, an expert on learning styles, has concentrated on four themes in his research into adult learning styles: 1) awareness by trainers of the need for a style of teaching different from that used with children; 2) the ineffective old thinking of training or learning theory; 3) the factors contributing to instructional effectiveness most commonly identified; and 4) the learner's perception of the qualities of successful trainers. His research indicated that learning styles affect the learner's ability to relate to the material, according to how it is presented.

Brookfield states that the adult's ideas about a good trainer affect their perception of the learning experience. I recall attending a class at the University of Washington where the teacher was late, rushed in, and began to read from the material we had already covered. I pulled out of the class. It is interesting to note that a friend of mine stayed in the class and said she learned a great deal.

According to Brookfield, our style is important to our ability to learn. How can a trainer provide training for all types? To fill the varied needs of all learners, the trainer can use an assortment of educational methods: demonstrations, quizzes, dialogue, questioning, visual based instruction, writing, problem solving, cooperative learning, debates, drama, peer teaching, simulation, trainee generated activities, and critical thinking exercises. This means expanding the current concept of console training.

Learning Styles

Expectations are at play in any training. Some students expect an organized approach to learning. For example; if I say there is going to be study time on Friday—there better be study time on Friday! Other students want to go with the flow. If we are working in the lab, having a great time, and a lot of learning is taking place, they will forego study time and fit it in later. In fact, they appreciate the dynamic learning as opposed to the organized. (Where do you think I fit?) There are often conflicts between styles. To gain cooperation from trainees, it is helpful to explain learning styles; and tell them you will attempt to use a variety of learning methods.

Like many of you, I don't put a lot of stock in standard surveys that tell you all about what type of person you are. But, I will say that the way you answer certain carefully formulated questions can say quite a bit about your thinking process. The 'Learning Style' surveys I have

Outstanding leaders go out of the way to boost the self-esteem of their personnel. If people believe in themselves, it's amazing what they can accomplish.

Sam Walton

used with my students have provided a great deal of insight into what they prefer in a learning setting.

What did I learn?

I learned to offer learning and training style surveys. The way you learn is valuable information and can contribute to making the training process rewarding and effortless. Make sure your trainers and learners are aware of the differences in learning and training styles; and that each is allowed to be flexible in their needs and methods. This is just another way for administration to show innovation and creative effort in developing a dynamic training program.

In addition to 'learning styles' there are also 'teaching styles.' Must the two match? Not necessarily. The trainer may be able adjust their style if they know what the learner expects. However, if the trainer has a very strong style, and the student has a strong opposite style, the two may not be compatible. If the trainer cannot adjust, it may be best to change trainers.

5. Adult learners must be comfortable and ready to learn

> *The physical environment and the physical and mental condition of the learner must be within a 'comfort zone.'*

Physical and mental distractions

Considering Maslow's Hierarchy of Needs, a trainee who is hungry, thirsty, cold, or uncomfortable will probably be unable to do the mental processing needed to learn. Physical distractions in the learning environment are a deterrent to thinking. Mental distractions can also cause a student to disconnect.

> *The U. S. Government Office of Personnel Management and the National Institutes of Health did extensive research on the time required to refocus on a task after an interruption. Researchers found that it took from five to twenty minutes for a person to reach a level of concentration where outside distractions and events were not noticed. After the interruption of a phone ringing, a question, a request, etc., it took an additional five to twenty minutes to reach that level again. Right dominants are extremely sensitive to the world around them and take longer to recoup than left dominants.*

People learn something every day and a lot of the time it's that what they learned the day before was wrong [especially in a Com Center—and with permission].

Vaughan

Similarly, if the student is distracted by life problems or personal discomfort or thoughts, it may be difficult to disconnect at all. ...Wonder and Donovan

Listening Problems

Since listening is vital to the learning process, blocks to listening are also blocks to learning. According to Wonder and Donovan, the following are listed as the most common blocks to listening:

1. *Physical: bad seats, too cold, too hot, can't see, have to go to the bathroom, ill, can't hear, tired, in a drugged state, odors, and other distractions.*
2. *Mental: poorly informed, too high a level of information, inability to concentrate. lack of self discipline, lack of interest.*
3. *Psychological: prejudice against the topic, bias against the speaker, 'I' centered and wanting to speak, low emotional state, mental illness.*

Hotels are notorious for climate problems. In a workshop in Houston I received great evaluations about the material and my presentation, but without exception, every person complained about the cold and the feedback from the mic.

6. *Adult learners cannot be forced to learn*

> *The learner must be active and willing to accept and work with the material and take responsibility for their own thoughts and interaction.*

Learner's Perception

The choices people make in their learning and the degree of effort they exert relative to those choices are influenced by many personal and environmental variables. I once delivered a stress workshop for an agency in two sessions, day and evening. The day shift group was resistant to the information, disagreeable, and clannish. Not an enjoyable teaching experience! I reluctantly dragged my exhausted, disappointed self to the next group. To my delight (and surprise), the second group was generous, involved, appreciative, connected. We had a great time! Same agency, same material. What happened? Group one had a problem with the Director, who told them I came to 'fix them.' They had closed their minds before I even said a word.

Few things in the world are more powerful than a positive push. A smile. A word of optimism and hope. A 'you can do it' when things are tough.

Richard M. De Vos

An adult can choose not to learn for a variety of reasons, even though the material may be relevant, needed, or valuable. It may be that the educator can do nothing to keep the learner from buying out of the learning experience. But often, the learner's withdrawal may be ameliorated by careful attention to the learner's needs.

7. *Adult learners' motivation to learn is related to their thoughts about their ability to succeed.*

> *Adults do not want to fail. If they perceive that their attempts at learning will not be successful—either from the trainer, from others, or by early perceived failure—they will lose motivation and desire.*

Discouraging the learner

In a college setting, students are nurtured. A good instructor is very motivated toward student success. The instructor is a guide, and if the student gets lost, the instructor tries to find them and get them back on the path. The relationship is frequently different in agency training. It seems that the agencies are trying to run trainees out. Then, if they hang on—they're worthy.

Adults *want* to be successful learners. If there is a problem with experiencing success, or even expecting success, motivation for learning may be detrimentally affected. There are many theories and research studies to support this. Some trainers believe that failure is motivating. For example, they grade harshly or are critical in order to encourage study. This is true for a few people who like challenge and competition, but for most, the opposite happens. The learner begins to question their ability to learn and therefore loses motivation to expend time and effort toward the learning.

Our agency trainers have a history of trying to scare the learners into learning. Hazing the learner—*"Let's see if we can get them to fail"*—is another form of discouragement. Total team support is needed for success!

Small successes as encouragement

Humans try to live up to other's expectations and want to believe the best. In my college classes I would often tell the class I expected everyone to score above 90% on a particular test because I believed in their knowledge on this subject. The results were always that 90% of the class received above a 90%! Of the 10% who did not, some would come into the office to explain. They never stated they

A person may not be as good as you tell her she is, but she'll try harder thereafter.

Anonymous

couldn't achieve 90%, they just didn't have time, or concentration, or whatever to achieve it on this test at this time.

In your agency training, build on small successes. Small successes add to the learner's self esteem and enhance their ability to take on new challenges and risks. A trainer can create small achievements and allow learners to build a foundation of accomplishment.

In a collaborative atmosphere, a trainee may also feel safe talking to the trainer about fear of failure, then opportunities for success can be designed. For example, I had a trainee with a fear of public speaking. Through negotiation, the trainee agreed to do a presentation while sitting at her desk. If the trainee believes the trainer is interested in their personal growth, the trainee may take on the fear in exchange for victory and personal development.

Conclusion

The trainer's involvement in the process of creating an atmosphere of success in the learning environment is vital. Although there are no certainties that an adult will gain the most of any training, there is a combination of factors that maximize efforts toward achievement. By training in a manner that shows the learner a concern for success, and establishing a safe atmosphere for learning, the degree of fulfillment for the trainee is highest.

Although trainers focus on the learner's success, the actualization of encouraging and inspiring another person to grow is exciting. To be a trainer means to crave that magical moment, that spark of recognition of the thrill of accomplishment. And, although the trainee owns this moment, it is not clutched tightly by the trainee, but is gratefully and proudly held out to be shared with the trainer—and that is very rewarding work.

Next—Great Creative Ideas!

Ten Creative Console Training Ideas

Ten Creative Console Training Ideas

In previous chapters we advocated using more than one way to learn. Now we thought it would be great to stimulate some ideas for you using adult learning theory.

Idea #1: Team Connection

> *Learners achieve better results in training when they are in an atmosphere of social acceptance, personal comfort, and team connection.*

Belonging to the organization and feeling a part of the entire team is a very important part of the learning process. If a person feels they can identify with the team, they are more likely to feel a sense of connection and comfort.

Create a 'yearbook' with a picture and bio of each employee, stating their hire date, job title, accomplishments, and whatever personal information they wish to share. A basic cut and paste document designed on the word processor is sufficient. Include a 'welcome' letter. This is also a team-building exercise that makes a statement—each person is a valuable addition to your agency. In turn, the new hire should be presented to the employees in the same manner (possibly by using a flyer). Include your agency philosophy, vision and mission statements, history, and current goals and projects. What a great way to build a team!

Idea #2: The First Week

> *True learning usually comes from personal discovery.*

Layout of the building and equipment is provided by a tour. This could be carried a step further by asking the trainees (as a group, if possible) to go on a 'discovery tour.' After their initial tour, provide them with a building layout map and a checklist of places and people they must seek out. [Hey! This sounds like a scavenger hunt!] Include the Fire and Police Stations, Emergency Rooms, ambulance

I prefer the folly of enthusiasm to the indifference of wisdom.

Anatole France

companies, and other contacts of interest. After they have completed the tour, the trainer can hold a discussion with them and answer any questions that were a result of their observations.

Idea #3: Nature Codes

Information related can be remembered.

Call types are often confusing to new hires who have had no criminal justice experience. Each person could be provided with a different set of call types to research. Some possibilities could be to 'compare and contrast' call types that seem similar. Another method is to read the actual law; i.e., assault; so they understand the need for proper classifications. It is important the trainees share and receive feedback from the trainer about their findings.

10 Codes

It is helpful to have a tape or transcript of officers and dispatchers using the codes and have the trainees see them used in context.

Idea #4: Phonetic Alphabet

More than one way.

It is helpful to have trainees practice spelling names with the phonetic alphabet by practicing first with their own names and then selected difficult names from the phone book or logs of names previously run. This practice should be audible—in other words they should be able to hear each other, dispatchers, and the trainers to hear what they should sound like when they have accomplished the skill.

Idea #5: Phone Exchanges and Locations

Group learning is fun.

To take the previously tedious task of learning area, exchange, and geographic areas to a higher degree of retention, students could 'own' an area and receive instruction to come up with some creative drill or presentation to assist the others in the group with names and numbers. They could wear the community names for a day, develop a rhyme or acronym for an area, or incorporate color.

The important thing to remember is that if you don't have that inspired enthusiasm that is contagious, whatever you do is also contagious.

Danny Cox

Idea #6: Call Taking Simulation

Modeling is necessary.

In order to learn, evaluate, interpret, and process the skills and thinking involved in call taking, the students must have the opportunity to practice in a simulated setting. A simple phone simulation setup is very worthwhile and can be provided with little cost or lab coordination. Once a call type is discussed and information gathering is 'modeled' by the trainer or a taped example, the students should receive a mock call. It would be a productive learning experience if the simulation lines could be recorded so students can hear their own call and evaluate their performance. Professional Pride offers a set of simulation exercises called *Ring, Ring: 911 Calls.*

Idea #7: CAD Lab

'Practice reality in steps.

Another suggestion from a trainee was to provide a more realistic lab session. Presently a caller sits beside the trainee and feeds a call to them. This is good for the beginning steps where the cohort gives them assistance, but then they should receive more realistic calls on a simulation phone, which they must enter into CAD. Another exercise is to listen to many calls and enter the information into CAD as it is heard on the tape. This is *real time* and adds some authenticity to their practice.

Many times the differing computer skill levels of the students is difficult to manage. When a trainee teaches, they learn, so it is worthwhile to ask the more advanced trainees to assist less experienced learners (with close attention by the trainer).

Idea #8: Videos, Readings

Expanded learning.

When asking a student to learn from watching a video, reading material (such as articles or the SOP), or even observation and riding, the students should be asked to 'do something.' Either return with an outline, questions, or a report. The trainer should then review and give feedback on the paper.

If you don't have enthusiasm, you don't have anything.

Kemmons Wilson, Sr.

Idea #9: Riding

Investigate, Think, Compare

When trainees ride with the police, EMS, or the fire department they should be required to fill out a feedback form that asks: 1) What I saw, 2) What I thought or felt, 3) What I learned, and 4) What I need to know more about. This should also be discussed with the trainer.

Idea #10: Scanner

Learning by exposure.

It was a suggestion from trainers to provide new hires with scanners. I believe this is an excellent idea. My students who had scanners picked up the radio faster. Students can then work with the trainer to understand what they heard and discuss how the call came in and how it sounded over the radio.

Dynamite Classroom Tips and Techniques

Remember back to a classroom teacher who annoyed you, lost you, or bored you. Then figure out a way not to do that.

Pivetta

Dynamite Classroom Techniques and Tips

One suggestion for improving your agency training is to provide classroom training. This can be accomplished in the agency, by combining small agencies, or by using your local colleges. Too often classroom training is perceived as lecture. NO! Lecture—that is, active lecture—can and should be a part of the classroom experience, but it never the TOTAL!

There is a revolution going on in our college systems about what higher level learning is. I recently took a masters in education course called 'Transforming Our Educational Systems.' New thinkers are asking that endless tiresome hours of lecture be taken out of the learning process.

Following are some unusual and creative techniques for teaching in the classroom that can greatly assist the learners in enhancing their critical thinking processes.

Questioning: When giving an active lecture, ask the class a question, then count ten slow seconds. If no one answers, call on someone and ask them for their thoughts.

Quizzes: Ask the learners to prepare a quiz from the material. This helps them to extract the meat of the information.

Tests: When giving a test it is important to make sure the learner understands the material. The old idea of surprising, tricking, or pushing the learner to study is false. Many times people downshift on a test and can't recall. The whole purpose of a test is to ensure the information has been learned. It can even be helpful to give the person the test to study. If it is comprehensive and you want the learner to know the material, why not? Retention comes from *understanding* the material, not just knowing it.

Teams: Create a team atmosphere by telling students they are a team. Give them projects or exercises that allow them to work together.

Lecture is telling them what you are going to tell them, telling them, and then telling them what you told them.

Anonymous

Name their group, give them an identity (Fall '96 Team, etc.). Team learning is dynamic!

Writing Assignments: Solid writing skills are important to this work. I would recommend daily writing assignments in their journals—either their choice or an assigned topic. The trainer should then read the journals and correct any spelling, punctuation, or grammar errors. Writing should be acknowledged in a positive way to promote trust in sharing thoughts.

Helpful Phrases: I recommend teaching customer service and crisis intervention techniques or phrases. A simple customer service video or Professional Pride's *Dangerous Opportunity* gives a lot of good information to the student.

Stress Management: An ongoing topic. Have the students search out articles from trade or other magazines on how to survive shift work, dealing with agency politics, maintaining self esteem, etc. Ask them to present their findings to the class.

Follow Up: During the classroom training, new hires will develop a feeling of team with their fellow trainees and classroom trainers. Have reunions occasionally. Schedule time when the entire group can get together. Provide a time when the primary trainer can meet one on one with each trainee. The trainee may not be able to be open in their work setting and valuable information will come from simply opening past lines of communication.

The Fine Art of Noticing

I am convinced that life is 10% what happens to me, and 90% how I react to it. And so it is with you…We are in charge of our attitudes.

Charles Swindoll

The Fine Art of Noticing

Somehow we never take into consideration that the new hire may need training on one of the most important aspects of this work—human relations. We just figure either they have good human relations skills or they don't. If they do the job is a match, if not, here again we have one of those nasty tempered dispatchers who can't seem to find a comfort level within the work, but otherwise does a good job.

The Human Element

Many of you have attended a Train the Trainer class, or CTO Training. There you learned to write an objective and to put together all elements of training and learning theory. Yet even after these skills are learned and practiced, training can still be dysfunctional. Many of our new hires are leaving their training unnecessarily (and it isn't because the trainer didn't know how to write an objective). The real exceptional trainer notices, listens, acts, and understands the basics of humanness. This understanding may be the one component that will put the golden touch on our training programs.

Losing trainees is turnover. Turnover costs money. Money that could be used for higher percentages in wage increases, better equipment, more personnel, better benefits. The human factor is ever present and it is a necessary part of the training process. Human relations is woven in and out of every aspect of this job, so it must do the same in the training program. Failing to look up and take notice, failing to take responsibility for improving the environment, or failing to recognize human relations weaknesses in your trainees will limit success.

The following material speaks to the human element in training—not the writing, planning, scheduling, or delivery—but the connection between the trainer and the learner.

Take a quantum leap forward into excellence and away from mediocrity!

Geese heading south fly in a "V" formation. Why? As each bird flaps its wings, it creates uplift for the bird immediately following it. By flying in a "V," the whole flock can fly at least 71% farther than if each bird flew on its own. Perhaps people who share a common direction can get where they are going quicker and easier if they cooperate.

Team Building
Peter Mears & Frank Voehl

Trainer Influence

Trainer's Attitude

When a trainer is assigned a trainee, is it possible for the trainer to suspend any pre-conceived ideas about who is and who isn't fit for this profession? (For example, religious people won't survive the 'crude' talk, outdoor people won't last inside, old people are too slow, young people don't have common sense, a beautiful woman will have problems with officers, men aren't as good, or men are better.) We all have beliefs like this, and they may have been shaped by real experiences or by personal opinion.

The funny thing about beliefs is: humans will manipulate every opportunity to confirm their beliefs. This doesn't leave much room for an objective approach to training! Even subconsciously we search out 'signs' to brace our theories. In addition, if we start out with a presupposition about a trainee, two things will happen: 1) the trainee will feel it, and 2) the trainer will fix on any confirming evidence and begin to build a case. Instead of reacting to an event, the trainer may over-react because a false or illusory case against the person has been formulated. Instead of one error standing on its own, it's placed high based on the pre-supposition, and then it gains much more ground.

The actuality about a hidden, pre-conceived belief is that it's not *hidden* at all. Adult learners are remarkably perceptive about their trainer's feelings toward them.

Trainer's Spirit

When you are with another person for eight to ten hours a day, it's impossible to hide feelings, emotions, and thoughts. If the trainer is having problems at work, is burned out, or is negative about the training process or the administration, it will rub off on the trainee.

> *I believe a trainee must get to know you, get to hear your deepest thoughts about your work, know where you came from, but.... But, if you don't like what you are doing—your job, or training—you shouldn't be doing it! It's that simple. This person has one chance at getting prepared for one of the most difficult professions; and they should have a guide that is strong, capable, and dedicated to that purpose. The trainee's success is your purpose.*

I am free of all prejudices,
I hate all people equally.

W. C. Fields

Training for this profession is difficult and takes a great deal of emotional and physical energy. You have to be up for it. So, what do you do if you are down, having a bad day, frustrated, angry, or whatever? Tell them. You don't have to go into detail, just let them know it's you, not them. You have every right to be human.

Group Acceptance

In each of my college training classes it was inevitable that there would be one outcast. It was obvious by the way the others would not listen or would automatically disagree with that person. There was tension in the air.

Another sure sign was the lack of support the person received from the group. If a person is accepted, the group wants the person to succeed. The team considers the person an equal, a valuable member. In an agency training program, how important is group acceptance to the successful training and retention of a qualified employee? And how important is group acceptance to the productivity and quality of work the employee produces?

We know the answer to 'how important.' Fundamentally important! Group sanction of a trainee means professional life or death. Group disapproval is not subtle and rarely goes unnoticed by anyone. Most of us have a very keen sense of when we are accepted and when we are not. A person recognizes even the most indirect, discreet feelings of disapproval. I spoke to an administrator recently who stated that two out of three of her trainees just weren't fitting in. This means, just weren't *accepted.*

The first step to noticing is understanding why rejection happens. If you understand human relations, you may begin to understand the team dynamics of 'approval.' The following is a discussion of differences that may affect a trainee-trainer or group connection.

Prejudice

Include the word 'diversity' in any training and automatically everyone expects white versus non-white. Those of us who have attended diversity training know race is a narrow part of diversity. Those who judge have all kinds of justifications for prejudice. Generally prejudice has to do with parental, societal, or isolated incident influence. Gordon W. Allport said, *"A prejudiced person will almost certainly claim he has sufficient warrant for his claims."*

The vast majority of human beings dislike and even actually dread all notions with which they are not familiar.... Hence it comes about that at their first appearance innovators have...always been derided as fools and mad [wo]men.

Aldous Huxley

Prejudice is a belief pattern a person *chooses*. If a person wishes to change their belief pattern, they will; but no *other* person can force a change of belief patterns. Prejudice becomes a problem when the belief patterns cause the person to act in a way that is harmful, discriminatory, or causes another person discomfort. We all know it is difficult to have a prejudice and not act upon it.

I recall a New Directions workshop in Fort Worth where I challenged people to think about their racial bias. One woman said, *"What if it's justified?"* My question to her was, *"An entire race of people hurt you?"* I asked her to isolate whatever incident or harm a person of another race did to her, to ask herself if her race is capable of the same action, and to begin to look around at the good people in that race—the ones who are just trying to make a living, raise their kids, go to church, get along. There are bad people in every group, but it isn't intelligent to judge the whole by its parts.

Lifestyle Differences

Besides skin color, there are many differences in people we can love to hate. There are differences in religion, beliefs, housing, education, leisure activities, sexual preference, dress, or family culture.

> *A pattern of reaction is the sum total of the ways we act in response to events, to words and to symbols...in their more obvious forms we call them prejudices. ...Hayakawa*

A closed group is defined as people with similar lifestyles, culture, religious beliefs, and possibly education. I have seen many Com Centers that were a closed group, and many that were an open mix! If a closed group is not accepting, it will be evident in intolerant remarks about other groups—blanket assumptions regarding certain types of people. It is important to recognize up front when a trainee has a different lifestyle that may affect the group. It's called the fine art of *noticing*. It is also important for leaders to be pro-active with employees. Provide diversity training, searching out beliefs that appear to be intolerant of dissimilar lifestyles.

Lifestyle rejection can go the other way too. The trainee may be not be accepting of a member (or members) of the group, or even their trainer. That judgment will show, and the team may reject the trainee's judgment in defense. An example was this snooty college graduate we hired who was very intolerant of smokers—nasty about

Behavior is the mirror in which each person displays their own image.

Goethe

it, in fact. Her obvious disgust for their choice to smoke led to her rejection by the group, even the non-smokers couldn't handle her.

You cannot change another person. If the differences with the trainee and the group are such that your training is affected, you will have to do something differently.

Appearance

Dolly Parton once said she couldn't help it, she always dressed like a hooker. *Should* we judge others by their appearance? No! *Do* we? Yes!

I recall a lateral from Seattle. She was a talented dispatcher, but she wouldn't wear a bra (and we're not talking about A cups)! There was nothing in our dress code that required her to wear a bra, but... . She lasted about six months. Another new person overdressed daily. That may have been overlooked, but many of her 'Sunday' dresses were way too tight for her very large frame. It was difficult to look at her and not wonder when the seams would pop and she would come pouring out. And then there was 'stinky,' a trainee who had mildew odor. It was difficult not to feel some resentment toward him, and no one was going to tell him.

> *We like people who are similar to us. This fact seems to hold true whether the similarity is in the area of opinions, personality traits, background, dress, or life style. Consequently, those who wish to be liked in order to increase our compliance can accomplish that purpose by appearing similar to us in any of a wide variety of ways. ...Robert Cialdini, Ph.D.,* The Psychology of Influence of Persuasion

Is it the trainer's responsibility to make the person likable by making them the same? Of course not! It's not possible. There are so many different types of people in most Com Centers. What I am asking is to be aware of the power of 'liking' because of similarity and 'disliking' anything different. Possibly you could head off a problem. Well, think of the alternative—putting hours of training into a person who eventually will not be accepted, or may feel rejection, or reject the group, end up in a personality conflict, and not understand why. Simply knowing what is going on is very healing to most difficult situations—a starting place.

If we had no failings ourselves we should not take so much pleasure in finding out those of others.

Rochefoucauld

Immaturity

Immaturity is like common sense, hard to define, but really easy to spot. We all love certain things about children, such as their spontaneity and love for fun. Children are also very self centered. It's natural to be self centered when you are young, but it's a terrible trait when a co-worker is consumed with ME. Often the immature worker is inconsiderate and unaware they must show interest in others. An immature person need not be young, just inexperienced in the ways of the workplace. Many lessons we learn about how to get along with other adults were learned the hard way—along the way. Some people just do not know how they affect others, yet with a few changes could become more likable.

The Know it All

"Everything I tell her she knows! It's a big argument every time I try to correct her." Many times a person feels the need to justify their actions when they are being trained. This may come from an unduly critical parent or teacher. If a person is harshly judged and often criticized, it is only natural to begin to develop a defensive posture. Of course this cannot be tolerated, but understand that this person has learned this behavior and must unlearn it. Many times the person isn't aware of the *results* of their responses. They may even think it boosts the other's confidence in their ability.

The result is that the trainer will often approach instruction cautiously. The trainer may feel hesitant to approach this porcupine. The educational effect is: it is difficult to learn when information is repelled back. (Be aware that people will challenge new information to what they already have in their knowledge banks. If it does not match, some people may seem confrontive about the knowledge, have a quizzical look, or ask further. This type of trainee behavior may be their natural way of learning and *not* 'know it all' behavior.)

Knowing the Rules

There are written rules and unwritten rules. Written rules are easy to know and the consequences are easy to measure. I don't necessarily mean only those written on paper, like the rules and regulations book (although trainee's must know and follow those too). I mean the obvious stuff people are expected to do in society, at work, in the world.

It may be an 'unwritten' rule that people don't blow smoke in someone's face, or swear in the workplace. However, as we both

Appearances are skin deep—don't judge a book by its cover. It's inside what counts.

My Mom

know, in Emergency Communications sometimes it is an unwritten rule that swearing is acceptable. (That's where I got *my* potty mouth!) This is a clear example of when an unwritten rule is not followed. It should be, but... . We justify it as stress relief.

OK. Then there are those written rules that are not followed such as breaks, arriving late, covering for each other, and sleeping on the console. (I recall a few bad nights catching flies. Did you see that guy on 20/20—national television—how embarrassing!) Trainee's respect first their peers and second the rules. They will not follow a rule not followed by those they respect—make sense? What does that mean to the trainee? *"Do what I say, and not what I do,"* does not work in an adult learning situation.

If there are written rules no one follows in your agency, bring it up at your next supervisory meeting. The rules needs to be re-visited. If the rule is found worthy, then it should be followed. If not, it should be discarded. If a trainee is required to act differently than directed, the result could be contempt for other rules, leading the trainee to survey other rules for possible leeway.

Unwritten rules are nebulous and blurry, but still have heavy consequences. Unwritten rules permeate every workplace, every home, and every classroom. Anytime people gather together regularly there are personality, power, and positioning issues in place. The fastest way to gain disapproval is to be unaware of, to contest, or to ignore the unwritten rules.

Seniority and territorial rights are two of the most common unwritten rules. Seniority brings privileges, written and unwritten; and the trainee is smart to understand this part of the workplace. Many times the immature worker does not recognize seniority as a reality and takes on a battle they are sure to lose.

One of my students called to say that Kate had messed with her unfairly and she was fighting back. I knew Kate. Well, let's see: Kate = 20 years of pounding rookies. Lynn = 0 years fighting with the likes of Kate. Kate could have been wrong, but I knew Lynn would be hitting the road soon. (She was let go one day short of her year probation.)

Beware of prejudices. They are like rats, and [wo]men's minds are like traps; prejudices get in easily, but it is doubtful if they ever get out.

Jeffrey

Territorial rights are another area of difficulty. It is very difficult as a new person to understand territorial rights. And it is very easy to step into someone else's space without knowing it. There is nothing a trainer can do when this happens but damage control. If a person wants to fit into any work circle they must be careful to watch, look, and listen.

When a new person comes into the picture, some people may feel threatened or uncomfortable with the change. Fear is their first response to the new—any new. They lack trust but don't really understand their own uncomfortable feelings. Many times my students were greeted with distrust. *("Those smart ass trained students. I learned it the hard way. I wonder what they know that I don't. It's not fair. Will they show me up?")*

The best way to stay out of territorial issues and avoid territorial trespassing is for *The New* to realize they have no 'territory' until they establish residency and that it is OK. Accept the reactions of others with non-judgment and let it pass. (It *will* pass.)

Negative Influences

When a trainee enters the new group, they may be barraged with either negative talk about the workplace, the administration, the contract, other employees, or officers. This type of negative atmosphere can create upheaval to the training process.

Another form of negativism may be directed towards the new hire. A feeling from the beginning that the person won't cut it. It may be somehow communicated that the new hire doesn't have a chance in hell of overcoming the rigors of the training. *"This is a really tough job; and we have lost better people than you!"* If the trainee has any problems at all, even those commonly met by even the best potential candidate, the negativism intensifies. A learner cannot experience peak learning if they do not believe they can achieve or accomplish what is ahead of them.

Negativism is expecting the worst, seeing all things as a threat to the present comfort level. Negativism is a first cousin to low self esteem. This could be personal or professional low self regard. Those who have been told they are not worthy—by low pay, poor care, inadequate training—begin to believe it despite their own resistance. A trainee picks up very quickly on a general attitude of lack of hope.

The habit of negativism can also be a learned behavior, a way to exert some opinion and get some attention. It's almost like a game to

Traits of a Negativist

...depressive look, rarely smiling.

...inability to accept any new idea as workable.

...little trust for those in control.

...harsh words for anyone opposing them.

...cynical attitude, but heart of gold.

...fearful and critical of change.

...display a helpless attitude at times.

... has a very healthy self concept at times.

...uses the same negative phrases repeatedly.

...wet blanket at anyone's success.

...gloomy outlook on life

...critical of those around them.

...seem to know there's a problem inside themselves.

...resistant to improving their attitude.

talk down and complain. It takes a strong person to break the cycle of negativism in a group. When a trainee or a trainer meets with group or individual negativism it makes it very difficult to stay connected to the task at hand. Negative talk and downshifted emotions are distracting to say the least.

The trainee may begin to develop a sense of doom. They worked hard to enter the profession, they were excited about the career, but now... . Maybe there is reason to doubt their choice and all this hard work may be for nothing. The trainee and trainer may feel a loss of enthusiasm and motivation—both essential to learning. If the new hire is bombarded with negative talk, their ability to learn is affected.

> *Trainees tend to believe what is said. When dispatchers typify a supervisor, officer, or new procedure as 'bad'—it may affect the trainee's future decisions and cause faulty judgments.*

New people may even be recruited into the circle of doom by troublemakers and complainers. If you notice your trainee picking up some negative influences, sit them down immediately and talk about it. Again, the fine art of *noticing.*

Living with Negativists

All professions have problems. One thing somewhat inherent in our profession is *Negativists.* A Negativist can really make your job difficult—as a peer, as an administrator, as a trainer, as a trainee.

Most of us have been negative at times, and most of us didn't like being that way. In this profession, we have all encountered the negative person. We don't like a negative person, but at times we mutely fall into step behind them. We seem to be so affected by them that they can drag an entire agency down. They are usually at the center of most dysfunction in the workplace, yet if you were to ask them, they are victims and only put forth their best effort—and more.

You will hear from a negative person some or all of the following:

We tried that. There's no sense wasting our time.

It'll never work.

THEY don't care about US.

What's the use.

This is a hard attitude to overcome when you are trying to make positive changes, or to plan, design, or implement something new.

Good…leaders create a vision, articulate the vision, passionately own the vision, and relentlessly drive it to completion.

Jack Welch

Why can't we just ignore them? Because there are parts of them in all of us. We've all been there. Usually this person is also an important part of the team—a hard worker, a person we value—so we tolerate their difficult nature. Realizing that a negative atmosphere is difficult to work in, what can be done to change these negative people? Nothing, and you know that if you know Rule Number One: you cannot change another person.

> *When I quit smoking, it wasn't because I didn't like smoking. I was feeling embarrassed about being a smoker, I didn't like smelling bad, and the cigarettes were getting too expensive. I was having more pain than pleasure out of smoking. Like most people, I like pleasure more than pain. To reinforce my decision I even gave the task more pleasure: I told everyone I was quitting and I took great pleasure in their praise. I have always had pride in my ability to achieve something. I saw quitting smoking as an achievement I wanted.*

The moral of the story applied to negativism is: if we want something to change, we have to want to change it, see it as painful, and that the change would bring us pleasure.

Understand the dynamic of the pain versus pleasure principle. What need is being fulfilled by the behavior? Does complaining give us pleasure, does it alleviate some pain? Complaining is part of a negative attitude, but there is more. Negativism is a way of internal reasoning that says *"I won't agree, cooperate, or trust until the world apologizes."* A workplace Negativist uses a lot of energy just wrestling with the feelings that direct their actions and affect their attitude. And their peers waste energy either by avoiding, joining, or disliking them.

> Occasionally, all of us feel a defeatist attitude—but we bounce back because we don't like to feel that way too long. Besides, most of us understand you can't win 'em all. Losing doesn't take away our entire being.
>
> Occasionally, all of us feel embittered about wrong bestowed upon us by others—but we forgive and forget. We understand we forgive others for ourselves.
>
> Occasionally, we are all disappointed because we had expected more out of something—but we get over it. Our spirit may be dampened, but we can shake it off.

To be valued, to know, even if only once in a while, that you can do a job well, is an absolutely marvelous feeling.

Barbara Walters

> Occasionally, we all feel unappreciated—but we find ways to get support. Mostly we can self talk ourselves into understanding that true appreciation comes from within.

Now, imagine feeling defeated, embittered, disappointed and unappreciated—but not having the ability to let it go. A Negativist believes they are defeated, because they tried and did not succeed, *and* it hurt. They expect defeat around every corner so they prepare for battle whenever they want something.

A Negativist feels embittered because they were wronged and it hurt, but they don't forgive because they learned somewhere that people had to *deserve* forgiveness. This person feels to forgive is to excuse or pardon, and forgiving would not follow the punish principle. Besides, when they are wronged the cycle of 'poor me' is strengthened.

A Negativist is often disappointed because they have unrealistic expectations of others and of themselves. A Negativist feels unappreciated and misunderstood—because they are!

To promote a negative attitude is to encourage depression. How can you avoid it? We all know to oppose a negative attitude is to feel the sting of a thousand killer bees. The first step is to recognize the person's feeling, but not the event. Acknowledge the person's feelings as real and valid, but say nothing to inspire expanded or extended negative feelings. Say, *"That must have hurt,"* as opposed to, *"That ass, he treats us all like that."*

A Negativist can cut solutions to ribbons with their razor reasoning and barbed memory of past failures. Involve the negative person in working on the problem—*not* the solution. Let them define the problem—that is their expertise. Remember, most Negativists got that way because they tried and failed. And they did try—once. What they need now is to trust the process so they can experience some success instead of failure again. Then they may be able to allow themselves to take just a little risk. It's a scary step.

What if you have a great idea, but become immobilized by negativism? How do you peel off a wet blanket? If a Negativist argues something won't work, acknowledge that it *may not*—but *may not* is not *will not*, and you would like to try anyway. Use the first person when proposing something. *"I would like to try this. It won't work? Well, I still would like to try, even though it may not work."*

Common sense is the knack of seeing things as they are, and doing things as they ought to be done.

Anonymous

Common Sense

If you were to ask any group of trainers what trait is most important to success in this field, their answer would be 'common sense.' Common sense is our 'divine omnipotent being.' We hold it up as a sacred trait, available only to those who were blessed with it at birth. It is believed you cannot teach common sense, therefore if a person shows a lack of this characteristic, the training program is essentially over and the person is washed out. I held this belief as a console trainer, but now I consider it a quandary.

As a trainer, it is my responsibility to determine every piece of knowledge, skill, and attitude needed to bring my trainee from zero to 100 in the shortest time possible. We learn in CTO programs about occupational analysis and listing all the job tasks, knowledge, and skills that fit together to make a person successful. We've done well breaking down the profession—expect for common sense. Common sense hangs out there: elusive, with no real definition, like some mystical *knack*.

Since we all believe a person cannot perform this work without common sense, there should be a measurement for this inborn quality. As a trainer I know if I want to develop a quality or skill in a person, I have to be certain what the quality looks like and how to identify it when I see it. It isn't practical to tell a person they do NOT have a certain quality for a job, and then, when asked to explain, have only vague answers about that quality. Shouldn't we be able to point out what happens when common sense is not used? What does a lack of common sense look like?

We know common sense on the phones and console means making the right decision or judgment, having instincts or even insight, without having to be told. It means figuring things out very quickly, it means trusting yourself to do what needs to be done—and doing it, usually right.

But wait! Now add the heavy duty stuff—without the benefit of much guidance or training. After all, isn't that how we know we have common sense: because we are able to 'wing it.' Isn't that what common sense is about—and confidence for that matter? In this game you cannot be trained for every conceivable situation. Situations have variables and those variables—well, vary—with every call.

So, we want to test a person to see if they can see things as they are, and do things as they ought to be done, and do it with confidence

Common sense hangs out there: elusive from any real definition, like some mystical knack.

Pivetta

under fire—without much guidance (like we did). We could give candidates a series of questions on a subject they know nothing about and see if they come up with the right answer most of the time. This subject haunts me. I worry about this. I have questions.

> *If we have to be born with common sense, then people with no common sense NEVER have common sense?*
>
> *Where do people with NO common sense work? Because I don't want to go there. If they have no common sense that means they don't see things as they are and don't do things as they ought to be done.*
>
> *How many people with no common sense are traveling 60 MPH next to me on the freeway? Yikes!*
>
> *If you don't have common sense, is it in just one area? I mean, can you have no common sense for dispatching, but common sense for computer programming?*
>
> ??? ??? ??? ??

I asked a friend of mine who sails the waters of Puget Sound if a person needed common sense to sail. *"Definitely! Without common sense you could get killed."*

I have common sense—I was a dispatcher—so I put myself on an imaginary boat, in imaginary water, to see if my common sense would get me through the unknowns. (Let it be known that I have never sailed.)

First, I would recall what is commonly known to all of us: wind pushes stuff. Then I could look at the sails and figure out what position they needed to be in so the wind could do its thing. Then I could appraise the equipment to determine how to get it to work.

How fast do you think this would all happen? Would I ever sail—or would I get killed out there just trying to make it work with common sense?

Now if I had a guide, I could learn about sailing and then show my common sense stuff by carefully following his guidance, listening and reacting instinctively, if needed. It's the *instinctual* reaction that is

Common sense is instinct and enough of it is genius.

Anonymous

common sense. When faced with an unknown, did I usually choose the right course of action?

I can vividly imagine getting yelled at, choosing the wrong rope, not knowing what to do—just standing there like an idiot. My sailing trainer would probably be shaking his head in disapproval, sure I would never make it. If I did not improve—and never could improve enough to speed up my reactions and choose the right rope—I probably would drop sailing.

We don't all have a propensity towards sailing—or dispatching. But we don't know until given every opportunity to learn, practice, and show our stuff. If a trainer does not let the learner make mistakes, react improperly, and look like an idiot, the trainer is not allowing learning to happen. If a trainee is pounced upon for every error, the capacity to take chances is lost. The trainee becomes hesitant and second guesses instincts. The common sense, immediate choice number one becomes tangled up with numbers two and three until what comes out is gibberish.

Instead of grabbing the rope and taking the chance of making a mistake, I would rather just stand there until told what to do. What comes out looks like no common sense when what it really is, is respect. I am respecting the difficulty of the task, and doubting my own ability to make the right choice.

If a trainer is not able to tolerate errors, the learner will not tolerate errors either. If the trainer is not supportive of mistakes as part of growth, the learner will not be supportive of mistakes as part of growth. What happens then is a downhill slide for both the trainer and the trainee.

The human side of training is understanding, asking questions, thinking how we think, challenging how we feel, knowing when we are right, knowing when we are wrong, knowing what we expect, getting what we need, taking nothing for granted, granting nothing for nothing, getting off our butts and getting educated, getting rid of our 'yeah-buts,' paying attention and noticing, being able to reach our butts with both hands, keeping our hands off other people's butts, but then there are some butts we can, but they aren't at work, anyway... . I must need some food!

Mistakes Are Road Signs!

How to Say It

Ways to tell people how you feel and what you observe in a manner that is not offensive.

Event—Effect—Change

1. State the problem or describe the event.

   "There are times when you don't show an interest in what the others have to say. Sometimes you talk more than you listen."

2. State the effect of the event or problem in your opinion.

   "With this group, people will stop listening to you if you don't take an interest in them."

3. State what you see that could be done differently.

   "I think you would really work your way into the team if you took some time to ask questions, listen, and remember little things about people."

I see—I feel—I need

1. State the problem, citing examples.

   *"**I see** many times when I tell you something you already know it or you question it."*

2. Relate the results of the behavior.

   *"**I feel** like I'm having difficulty training with you."*

3. Ask for assistance in solving this problem.

   *"**I need** to understand when this happens why it makes me uncomfortable and what you mean by your defense."*

4. Be clear on your expectation of future behavior.

   *"**Would you** work with me to recognize when this happens so we can discuss it at that time?"*

If the behavior continues:

Repeat steps 1 and 2 above, then continue as follows:

3. Remind the employee of the earlier talk.

   "We talked on (date) and I said then that...."

4. Cover the results of the continuing behavior.

   "I feel now that...."

5. Offer a timeline for change.

   "I need...."

6. Provide an overview of the meeting in writing.

   "We will..., and then...."

Real Real Wrong

Honest criticism is hard to take, particularly from a relative, friend, acquaintance, stranger, or co-worker.

Jones

Real Real Wrong

OK, you can't save everyone. There, I've said it! The Exceptional Trainer can and certainly must know when a trainee isn't going to make it. There are many reasons why people don't make it through this training—or any on-the-job training. In working with hundreds of students, I conducted my own internal assessment and category lists of those who wouldn't make it. I learned quickly I could not identify those people the first day or even the first week. In fact, if I did try to identify a 'winner' or 'loser' I was usually wrong. But there did come a time when I realized a person was not meant for this work, or a person did not have what it takes to learn (or do) this job.

I could tell from the person's

...inability to go past a certain point
...confusion that persisted
...slowness in the essential skills
...lack of growth in understanding
...inability to take risks
...fear of failure
...resistance to new information
...look of misery on their face!

You will know when your trainee has reached that no-return point. You are an expert at this work; you can feel, see, hear, and instinctively know when something isn't right—and if it isn't going to get any better.

Not everyone likes or is fit for this job. I certainly could never be a police officer. You would be able to tell by my disinterest in doing things the way they wanted me to. Imagine yourself in a profession in which you know you don't belong. Now, what would you act like after about six weeks of training? Confused? Irritated? Lost? The problem with some people is they have a romantic idea about this job, (like any Emergency Service) and they want desperately to succeed. They have told their family and friends, and are proud of their new position. To let go is to admit failure, so they hang on, looking for excuses why they don't live up to training standards.

Often there are disputes regarding letting a trainee go, and the trainer must justify that decision. The best method for letting a person go

He who has burned his mouth, blows his soup.

Anonymous

without dispute, is to develop an atmosphere of trust and communication from the beginning. If there is a system of feedback, of checks, and of ongoing documentation—clear, honest, and open—there is no room for miscommunication. If the trainee believes you have their interests at heart, they will trust your leadership. And, if you have feedback forms, check sheets, lesson plans, and evaluations in place there should be no question about the action taken.

If you are unsure, or are challenged regarding letting a trainee go, you may want to bring in an outside evaluator. Have the evaluator sit with and observe the trainee. The only information provided should be a copy of the training schedule, and the knowledge check sheets. No evaluations, feedback forms, or any other information obscuring the view. This outside evaluator will be able to look at the SKILLS and not at the PERSON. The assessment from this evaluator should be presented in a very specific format. For example, the evaluator should have a set of questions to answer regarding what they observed. If your evaluation forms offer a view of what skills should look like at this point in training, the new evaluator can assess if they do.

Trust your knowledge of the *work* and the trainee's *ability* to do the work. Too often we are concerned about too many other things. Too often our judgment is not valued because we are not prepared to back up our evaluation. Would you think too many poor candidates are kept, or too many good candidates get away? If our training programs involve constant communication both ways, a good evaluation system, and trainers who pay attention to learning theory, we will probably err by keeping too many bad candidates. A good candidate will not only survive but thrive in this atmosphere of functional higher level learning opportunities.

THE VISIONARY TRAINER

Creativity is to see what everybody else has seen, and to think what nobody else has thought.

Gyorgyl

The Visionary Trainer

When we are involved in learning, we become a part of something very exciting. Both the leader and the learner can harvest great rewards. However, I will acknowledge that the training process can be frustrating, draining, and difficult. The trainer may face difficult learners, lack of resources, too many demands, poor leadership. Being a trainer is not for everyone. But, if taking on the challenge of being a learning facilitator appeals to you, take caution, because it gets in your blood and can become all consuming!

In this chapter, I want like to discuss those people in this industry who are taking the leadership challenge today. We'll call them 'Visionary Trainers' because they can see ahead to the possibilities. I know them because I pay attention to what is going on in this industry. I know them because they contact me. A visionary will seek out knowledge and support. Today's trainers are faced with a multitude of immediate difficulties, but they cannot be stopped. Here is a brief snapshot of what I see when a Visionary Trainer appears to me.

Visionary Trainers see training with a larger focus, not just as the person who takes a new hire through the process, or the person who works on agency training. They take training out of their own realm and push the limits and boundaries set up for them at every opportunity. They have stopped asking permission and occasionally have to ask forgiveness. They are politicians and diplomats, resourceful and hard working.

The Visionary Trainers are involved in classroom training.

They understand that classroom training is a hands-on affair where learners have an opportunity to discover, explore, bond, compare and contrast, and have fun while learning to think like their trainers want them to. These visionaries develop their own skills at managing a classroom setting. They learn to enjoy the setting. They know that real learning is going on if the students are doing most of the talking. The classroom needs to be a noisy, happy, and challenging setting.

They take agency classroom training a step further. Maybe the agency does not offer classroom training; for example, the center is too small, or the administration does not buy off on it. If the trainer knows the worth of this interactive type of learning, the trainer will single-

The vision of things to be done may come a long time before the way to do them becomes clear, but woe to him who distrusts the vision.

Jenkin L. Jones

handedly pull together an opportunity for classroom learning and also offer it to others in the area who are interested. This is a spirited 'build it and they will come' attitude. The visionary trainers that I am in contact with go outside their agencies to put on training—sometimes at their own expense. (You know that I am outspoken about the need to pay our trainers, however, sometimes the rewards for carrying the expense—or training for free—come back in the form of career rewards.)

Now, let's really take the classroom training thing way out there. There should be a college training program for 911 in every city or town that has a college offering trade or occupational training. The reality is that there are only a handful of college programs in the United States and Canada. Why isn't this happening?

We need visionary leaders who can take on the additional encumbrance. It's not just that forming a college program is a time-consuming process, it's a step into uncharted territory. Remember, most people fear doing something different. Visionary trainers acknowledge the need for training in our profession at the college level. They are reaching out to the local criminal justice training programs and other institutional learning environments, but they understand the need for connections and support from the academic community as well.

The Visionary Trainers are sharing what they have learned.

I worked in a Com Center, but I have come light years from those days. Sometimes I have to remind myself that putting these great ideas to work isn't all that easy. Visionary Trainers meet with attitudes and all kinds of roadblocks. I can share what I know, what I have learned, and what my vision is, *but* I am not in the trenches. Visionaries are out there, and when something works, they share their successes and struggles with their associates.

Here are some examples of what Visionaries are sharing:

- *How to motivate those who are forced into mandatory learning.*
- *How to work with learners who take time, when time is not available.*
- *How to put together justification for an increase in funding for training.*
- *How to find alternative funds for training.*

The people who shape our lives and our cultures have the ability to communicate a vision or a quest or a joy or a mission.

Anthony Robbins

- *How to put together no-cost learning.*
- *How to sell the administration on new training ideas.*
- *How to find and set up a simulator for training.*
- *How to be creative on console training.*

Visionaries are the ones who write articles and books, or just stand up at APCO or NENA meetings and speak out. They are the ones who formulate mentoring groups, or trainer's support groups in their areas. They know the fine art of networking. These are the people you see at APCO and NENA meetings sitting quietly at first until they learn the system. Gradually they become involved and get their name out there as a leader.

Visionaries are self-directed learners.

They never stop gathering new ideas, thoughts, inspiration, and theory. Their own learning even turns them on. They read books, go to classes, read industry magazines, talk to people, call up people like me and pick my brain. They reach out and reach within. They put true spirit into their work as a trainer. They know how to say, *"I don't know,"* then they quickly add, *"but I'll find out."* Or better yet, they say to a trainee, *"I don't know, but why don't you research that and share what you find with me."*

And most important, Visionaries challenge the status quo.

No matter how great something is, it can be improved upon. In Peter Senge's book *The Fifth Discipline*, he calls this the 'constant attention to improvement.' Visionaries practice the fine art of noticing—listening which turns to action. Their action may be nothing more than a spark of a thought. Remember that thoughts are things, and every creation began with one thought.

You may be able to recognize one of these Visionaries. They may be satisfied, but they are never content. They may be exhilarated, but never comfortable. They may be pleased with their progress, but never smug. And, at this point you can recognize a Visionary if you just look in the mirror. *You* have the ability to create and innovate, to become a leader, and an originator. *You* are the architect of your own career. *You* are the cornerstone for the building of a healthier training environment for your industry!

Do You Love Appraisals? You Should!

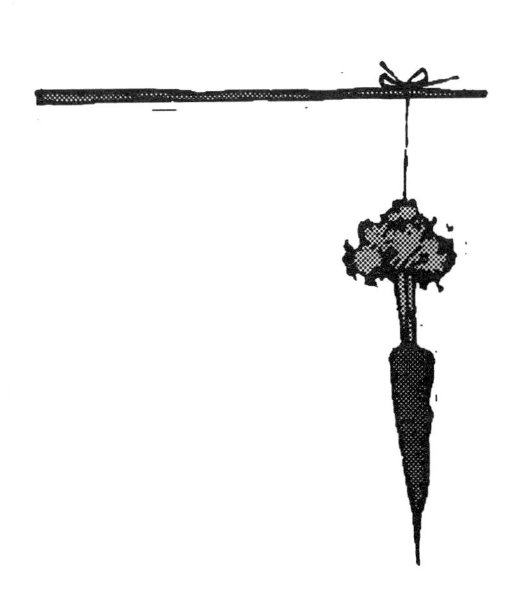

EVALUATE:

1. *To ascertain or fix the value or worth of.*
2. *To examine and judge carefully; appraise.*

The American Heritage Dictionary of the English Language

DO YOU LOVE EVALUATIONS? YOU SHOULD!

> **Resolution**
>
> ***Let's get this evaluation thing down once and for all. Take a serious and hard look at your evaluation process. Do you feel good about your form as a learning tool? Do you feel your employees are motivated by their evaluations? Are you? Although evaluation programs take some effort, they really are at the heart of the work—and we need to take care of our heart to keep it pumping good revitalizing energy through our agencies.***

One of the most challenging functions of a trainer, supervisor, or administrator is providing evaluations. It's never easy for either the person receiving the evaluation or the person giving it. If the evaluator is unskilled at one-to-one communication or uninformed about the true value of an evaluation, the results can be disappointing.

> *I was apprehensive because I had never had an evaluation from this supervisor. I didn't know what to expect. I was hoping there would be some rapport, some feedback on how he saw me. (I was a little anxious coming from another agency.) We began by talking about some light stuff, he handed me the evaluation form, asked me to read it and get back to him if I had any questions. That was it. The scores were all high and there weren't any comments. I was a little disappointed, but no harm.*

Here you can see that the employee viewed the evaluation process as an opportunity to get and give feedback, which didn't happen. Although the employee may have felt OK with the ratings on the

When was the last time you had an evaluation that was useful, a real learning experience?

Pivetta

form, the supervisor missed an opportunity to connect, motivate, and listen to the employee. This supervisor took it for granted that the employee was doing well, and he did not need to set an atmosphere of communication where any problems would have surfaced. In a job like Emergency Communications, a worker is independent and much can happen without the supervisor knowing. The worker may even have insecurities, but no avenue to address them with their superiors.

The time spent doing an ineffective evaluation can be wasted as in the above example, but no harm is done as the employee says. But what about when harm is done?

> *Last year I was fine, got good marks, and even had a letter from a citizen attached. This year I had 'needs improvement' from the same supervisor. I was shocked. I hadn't had any problems that I knew of. I didn't understand, and I asked why. She said she had been rating everyone too high in past years, and this better represents a fair evaluation.*

Here the employee has a sense of, 'No fair.' It is estimated the majority of untrained evaluators rate people too high, wanting to motivate and inspire their workers. We all want to be liked, we all hate giving bad news—evaluators are no different. The problem in this example was the process used to change the evaluation ratings and the resulting incitement of discouragement. There is no way to take this back or fix it.

Evaluations are not valuable if the evaluator does not provide adequate time for assessment. Evaluations are not valuable if the person evaluated does not trust in the evaluator's process of assessment.

Evaluations can be difficult because they have to do with emotions, both for the evaluator and the person receiving the evaluation. By combining the Evaluation Program Goals with a forthright recognition of problems, the evaluation process begins the change mechanism.

In order to have effective evaluations you need several components: a trained evaluator, an evaluation process, and an evaluation tool.

Why Do We Evaluate?

Evaluations measure performance against a standard. Evaluations also ensure the work force is indeed able to carry out their job in the manner expected. Evaluations protect the agency against liability, and

I repeat: *Evaluations should only be offered after thorough assessment.*

Pivetta

can be a great asset in promoting, training, and motivation. Even though your agency may have an evaluation process that causes morale problems, remember that adults really do need feedback on how they are doing. Not just good feedback, but authenticate answers to the question, *'How am I doing?'*

How Do You Design an Evaluation Process?

Designing an effective evaluation program for your agency can be initially time consuming, but once in place the difference in how your people feel is outstanding. A total appraisal system should be solid and reliable—producing results that both agency and personnel need.

It is important to first 'define' levels of performance—this does *not* mean just listing what needs to be done. Next, determine who should measure and how often evaluations are most serviceable. Finally, appraisers should be trained in the new system and all personnel 'sold' on the process. It is important to know that an evaluation program is a dynamic, living, moving, and changing process. How long has it been since you scrutinized your evaluation program?

Beware the trainee who is afraid to make mistakes.

Pivetta

10 Evaluation Benefits

1) **Protects Agency from liability.** Evaluations are used in court both for and against the employee.

2) **Provides accountability for all sides.** In an honest evaluation atmosphere, supervisors pay attention to the employee's work performance, and employees strive to perform to expectations.

3) **Provides adults with the needed feedback.** Adult trainers, educators, and psychologists all agree about the importance of honest, timely feedback. This is a basic human need. Adults will ask for feedback if it is not received.

4) **Allows the organization to clarify or emphasize agency objectives and standards.** How else can needs be identified than in candid assessment?

5) **Evaluates individual Call Receiver's or Dispatcher's needs and strengths.** This is especially necessary in Emergency Communications where workers are self directed and perform their job without direct scrutiny. The work is difficult to measure without direct assessment.

6) **Identifies training needs and employee problems.** By providing Telecommunicators a vehicle to communicate needs or problems in a safe atmosphere, the agency is empowering workers and offering a realistic overview of agency functioning.

7) **Provides an opportunity to motivate, support, and appreciate individual contributions.** We often say this is a thankless job. What better opportunity to say, *"Thanks for a job well done"?*

8) **Provides administration with a tool for promotion, employee development, and coaching.** A truly well run Com Center will recognize and develop potential leaders through the evaluation process.

9) **Enables Administration or Supervisors to pass on workplace goals or changes.** Another chance to implement needed changes and have the employees buy into a process by making one-on-one contact. This makes each individual responsible for the success of the agency by communicating.

10) **Builds groundwork for trust and honesty.** There is so little opportunity to communicate. This is great one-on-one time.

What hinders creativity? Fear of making mistakes, fear of being seen as a fool, fear of being criticized, fear of disturbing tradition, fear of losing the love of the group, fear of really being an individual, fear of being rejected, fear of not doing a good job, fear of failure....

Koberg

11 Fundamental Truths about Evaluations

Or The Tao of Evaluations! or Zen and the Art of Evaluations. Here are some thoughts about the evaluation and assessment process. Look carefully at your agency process and determine if your organization fills these needs.

1) **Employees need and want feedback on how they are doing.** If an evaluation is not carried out in a regular and timely fashion, employees can become unsure and/or unmotivated. They may actually be participating in behaviors and practices that are incorrect, or even dangerous. If an employee is not offered feedback, the assumption is, "You and your work are not of value." Many agencies do not provide time for trainers to provide proper evaluations.

2) **Evaluations can cause harm if not valid.** Feedback is NOT useful if the information: a) does not come from a credible source, b) if the process is not timely, and c) if the information conflicts with what the person being evaluated already believes.

3) **Evaluations should ONLY be offered after assessment.** What is your assessment and validation process? Do you provide time for your trainers and supervisors to really analyze the skills, knowledge, attitude, and growth of each individual?

4) **The use of tapes as examples is vital.** If a carpenter were to be evaluated you would look at their work. Keep an archive or portfolio of each Telecommunicator's work.

5) **Self evaluations are essential to growth and understanding.** If an agency is not providing a method of self evaluation, a very important learning process is overlooked. Not only is it valuable for the relationship between the evaluator and the worker to scrutinize the work—it is important that both agree with the assessment. Change will not take place if the person believes there is no need to change. Motivation will not be there if the person believes the source is not credible. The evaluator and the employee should come out of a session on parallel paths.

6) **More than one person can provide feedback on a person's performance.** Having two or more evaluators adds validity to the process. Given the criteria, the assessment tool, and the

I hope someday to have so much of what the world calls success, that people will ask me, 'What's your secret?' and I will tell them, 'I just get up again when I fall down.'

Paul Harvey

actual examples of work done, ideally both should come up with the same assessment.

Also, we evaluate others based on our own ability. Dual evaluations can help eliminate this. Dual assessment is especially useful: a) when there is a difficult call, b) to support praise, or c) when an evaluator admits they may feel biased.

7) **Extreme ratings must require proof.** If the scale is 1–5, where 1 is unacceptable and 5 is superior, both a 1 rating and a 5 must come with either a tape, a written complaint, a written commendation, or the like.

8) **Number ratings should be defined.** If a 1 is Unacceptable what does that mean? What will happen? Are you...? Then what? If the next is 2 what does that mean? Unacceptable, but not deadly? One up from dog meat? Eliminate blurry ratings.

9) **Rating descriptions should be offered.** Let's say you have a rating called: *Able to deal with upset or angry callers.* What does it mean to be able to 'deal with' an upset or angry caller. What does that look like? What does it look like to NOT be able to deal with an upset or angry caller? Maybe an unacceptable (or 1 rating) would be described as: *Argues, debates, or attempts to control and 'win' in situations; or resists callers who are angry or upset.* The evaluation form is now a learning tool. This rating, accompanied by a tape, is defensible in court and to the person receiving the evaluation.

10) **Knowledge, attitude, and skills should all be measured.** Measure only *observable* attitudes. Attitudes result in behavior.

11) **Goals should be attached to the evaluation form and process.** After the evaluation is completed and both parties agree on the ratings, attach goals on which both agree. Include training, personal achievement, improvement, acceptance of past poor performance, and an understanding about what to do differently or continue to perform well.

Service is just a day in, day out, ongoing, never ending, unremitting, persevering, compassionate type of activity.

Leon Gorman

12 Common Problems With Evaluations

1) **No set objectives for Supervisors to follow when providing evaluations.** Trainers or Supervisors are not sure what is expected of them in the process of evaluating.

2) **Supervisors are not provided with training on how to communicate effectively in an evaluation setting.** Again, most of us do not want to point out faults, bring up unpleasant events, or discuss difficult issues. In an honest atmosphere, where support and growth are emphasized, the evaluator can let go of that feeling.

3) **Supervisors are not provided with enough time to adequately measure performance.** Listening to tapes, interviewing field personnel, following up on information about performance, filling out evaluation forms, preparing paperwork, and providing one-on-one sessions are all very time consuming.

4) **Supervisors may use evaluations inappropriately.** The common 'old' practices of using subjective criteria and personal feelings invalidate results.

5) **Supervisors are not consistent from shift to shift.** The day shift trainer rates high, the grave shift trainer rates low.

6) **Working Supervisors are put into an uncomfortable middle position.** The feelings of loyalty to both administration and crew result in the evaluator taking the 'middle road' on evaluations.

7) **Telecommunicators do not consider the evaluation process as valuable to them.** This attitude is the direct result of not believing the evaluator has done their homework. Invalidating evaluations is especially true when the person being evaluated has a different view of their performance and abilities. Often evidence is not produced to support the ratings.

8) **Supervisors are evaluated by administrators who do not have an effective measurement tool or process.** Supervisors do not have a proper evaluation tool.

9) **Tenured employees and trainees are rated on the same forms.** The knowledge and skills expectations are different for tenured or veteran employees than for trainees. The measurement tool must reflect that difference.

It's too late to agree with me—I've already changed my mind.

Anonymous

10) **The mood for evaluations is one of obligation instead of opportunity.** Of course, without a well thought out evaluation process, the time spent is perceived as *obligation* and not *opportunity*. Supervisors are expected to 'find' time to evaluate, employees are rushed through the process before or after shift. The trainer may be seen as resistant or procrastinating, causing much distress. Human nature says, *"There must be a problem."*

11) **The evaluation process is unproductive, wasted time.** Everyone is worried about their own interests and unclear how to state their needs without causing trouble. People don't communicate needs either way, or it's a one-way meeting. Either or both parties leave feeling discontented.

12) **Evaluations have been used to control, manipulate, or punish employees.** The result is ambivalence or resistance to the whole process. The information is seen as not only useless, but destructive.

In a Rut?

I grew up in Montana where the snow would get piled so deep in the main streets that you would get your car into a rut in the road to travel along. If you wanted to go anywhere but where everyone else went, or get onto a side street, you had to plan ahead, slow down, then aim. It was hard work to get out of the rut, slipping and slding, bumping and struggling. If you didn't make it, you'd try again at the next opportunity. If you didn't make it out at the next road, you'd end up wherever the road took you.

Pivetta

What Makes an Exceptional Evaluation Program?

Flexible apprentice evaluation forms that reflect the actual immediate training provided.

Benchmarks should be provided for training and evaluations that mirror training during a specific time period. In other words, in the last month, what was provided in the area of knowledge? What was the measure that the knowledge was received? A written test? Did the trainer teach to the test? Then, did the trainee retain and understand the knowledge imparted?

> *Example: The trainee is required to become familiar with the names of all the city officials in the police department. This is knowledge. The measure of the achievement of this knowledge is an inquiry, either verbal or written. The knowledge is gained or it is not. The trainee has 80% of the names down. Is that acceptable at this point?*

What was provided in the area of skills training? Did the trainee have sign off sheets and the ability to identify when the skill was instructed, performed, and evaluated? Was the trainee then provided with the chance to demonstrate the skill being rated?

> *Example: The trainer demonstrates how to handle a transfer call. The trainee is then observed handling a transfer call. Did the trainee perform all the steps necessary to carry out the task? If certain steps were missed, was the trainee notified and given a chance to correct the error? Did the trainee have the chance to then demonstrate the skill again, after the correction? The employee would receive a satisfactory only after completing the task successfully several times.*

How about attitude? Were the required job attitudes clearly taught, and did the trainee provide feedback that those attitudes were understood? Were attitudes outlined in direct correlation to observable behavior? Did the trainee then exhibit compliance to and willingness toward these expected attitudes?

Low ratings are goals for the next rating period.

Pivetta

> *Example: The trainee has signed off on rules and regulations regarding arriving prepared to work ten minutes prior to shift. If the trainee consistently follows these rules, the evaluation should reflect adherence.*

Advanced employee evaluation forms that measure the actual responsibilities, duties, knowledge, skill level, and attitude expected.

Once a worker passes probation, it is expected that job duties are performed to an acceptable level. First, the worker's job description should reflect *all* of the duties, tasks, responsibilities, and assignments required during the shift.

Next, the job performance evaluation should reflect ONLY these components. In other words, if it isn't in the job description, it should not be evaluated. And if it is performed, but not in the job description, the job description should be updated or the employee should not be rated on this aspect.

Finally, the evaluation, containing all aspects of the job requirements, should embrace how well the employee is expected to perform to meet the agency expectations.

> *Example: The employee is expected to answer a 911 line in three rings. This is expected. If the call receiver does answer most calls in three rings they meet expectations. If the call receiver answers the calls in one to two rings (given a busy shift) the employee may be rated as 'exceeds' expectations.*

Too often in our attempts to provide positive feedback to employees in this difficult, thankless profession, we get ourselves stuck in a tar baby. We evaluate people who 'meet' agency expectations as 'superior.' This sets the tone so that a 'meets' is considered inferior.

> *One of my co-workers complained,* "I am not just average. I work hard to rise above average. I come in early, I stay late, I miss lunch, and I work hard to always be pleasant to be around." *There was no language available for the supervisor to outline what constituted a superior employee. Nowhere did it say:*

Unhappiness is in not knowing what we want and killing ourselves to get it.

Don Herold

'Should the employee come in 30 minutes early, stay 30 minutes late, miss one out of two lunches, and remain consistently pleasant to all persons encountered, they shall be considered far superior to others.'

This was this employee's interpretation of what a superior employee did. My suggestion was to go back to the Administrator and communicate the need to exceed expectations. Ask the Administrator to identify what must be done to receive a superior rating. If the employer is unable to identify what a superior employee looks like, they cannot expect their employees to put out the extra effort required to improve.

Both the evaluator who never gives a 'superior' and the evaluator who is generous with 'superior' are ineffective and create morale problems. The rating system will be suspect and ineffective if those being rated do not feel the ratings are deserved or realistic. If doing what you are supposed to do is superior, it doesn't leave much room for those who really do consistently perform exceptionally. Yet in a job where meeting expectation is difficult anyway, how do you provide motivational feedback to your workers? Truth is, no person wants to be average if they are working hard at a difficult job.

One way to qualify those who 'meet' and those who 'exceed' is to include the conditions under which the employee performed. Sure they answered all calls within three rings, but there were only 200 calls in four hours. Another way to raise a 'meets' to an 'exceeds' is to notice that the person unfailingly complied over a long period of time or exhibited long-term sustained performance. This is worthy of note.

If you already have a problem with the rating scale on your evaluation forms, start over with employee input. Ask the workers how they would want to be rated when they do what is expected—when what is expected is plenty. Appoint or request a cluster of line Telecommunicators who can discuss the concept of ratings and come up with a fair and equitable system that provides motivation—and reality—with room for superior performance.

Time provided for appropriate assessment of employees.

If the trainee is closely monitored, the trainer has the opportunity to gather and provide examples to support the ratings. The trainer can even earmark tapes for the trainee's listening pleasure.

If you could only get across to our Com Center trainers what they are doing to these trainees. I hired a woman of great potential last spring. Unfortunately she ended up with a trainer who was known to be harsh. I came back from vacation and the trainee was gone. They said she quit. I was upset because I knew this trainer had run her out of here. Anyone with a brain would not want to work where their first impression is that they aren't really wanted.

Deputy Director

However, if the person is working independently, it takes real time to listen, and listen, and then evaluate. The calls must be matched to the information provided in the call. This takes time. As unrealistic as it seems to provide time for supervisors or trainers to listen to each employee's tapes, this is the only way to honestly find examples of good work, or substandard performance. Some supervisors evaluate by the 'feel method.' (Kinda like how I do my checkbook.) Times have changed and so should your evaluation procedures.

Self evaluation and goal setting.

An evaluation 'process' allows for the employee to agree with the evaluation so that it is valid. Self evaluation, given clear guidelines, allows the learner to spend some time agonizing over the same questions as the trainer. It's not so easy when you are asked to actually write it down. If there are disagreements, the trainee is allowed to provide examples or justification of their own work. That is assuming the employee rates categories higher than the trainer. If the trainee feels their performance is *lower* than the evaluator's, it is important to recognize the trainee's feelings of inadequacy and discuss expectations and levels of performance. Maybe the trainee knows something the trainer doesn't know?

Goal setting isn't New Age, it's a great management tool. If an employee is doing adequate work and desires to do superior work, the trainer-trainee team can discuss how to increase skills, knowledge, and performance. Setting goals also allows the trainer a foundation for the next evaluation for positive feedback or documentation for lack of performance.

Consistency in the time evaluations are provided.

When should an evaluation be given? Right after assessment! However, conceding the evaluation process is weighty on an already busy supervisor or trainer, evaluations are the soul of quality control, motivation, liability, and agency excellence. How often should assessment happen? During the early training stages, evaluation should be quick and easy to do. Providing training guidelines and free form daily evaluation sheets are ways to expedite the process. The trainee should have immediate feedback but only occasional evaluations, perhaps weekly at first, then expanding to monthly.

After a trainee is released from training, it is essential for the trainer to keep tabs on the trainee, making regular visits to listen to tapes or talk to coworkers or field units. This activity doesn't need to be overt or

We live by encouragement, and we die without it—slowly, sadly, angrily.

Celeste Holm

covert, just a part of the supervisor's duties. This attention is appreciated by the trainee, and the evaluation has much more validity for both the agency and the trainee.

The veteran Telecommunicator should be provided with feedback at least yearly, but it must be a valid report. A supervisor's job of evaluating veteran employees should be ongoing. When a great call is noticed, write it down. When a problem happens, write it down, too. It's easy to let both get away without notice. It is also helpful to listen to tapes to check normal everyday skills and attitude.

Inquiry & Investigation
Review Sheet

To: Trainee
From: Trainer
Re: Incident Investigation

This call has come to my attention for investigation. This call, dated ___________, was handled by you. I would like you to participate in the inquiry process by completing the following steps.

1. Please review the attached procedure regarding this call type.

2. Review the enclosed tape (or written description of the event).

3. Attach your observations of the incident as you recall it.

4. Provide a written comparison of the procedure to the event

5. Write a conclusion regarding your actions and the call.

6. Sign and date—return to me by: _______________

Signed by________________________ Date: _______

Received by: ____

cc:

Console Reading

First make sure that what you aspire to accomplish is worth accomplishing, and then throw your whole vitality into it. What's worth doing is worth doing well. And to do anything well, whether it be typing a letter or drawing up an agreement involving millions, we must give not only our hands to the doing of it, but our brains, our enthusiasm, the best—and all that is in us. The task to which you dedicate yourself can never become a drudgery.

B. C. Forbes

CONSOLE READING

Active Learning, Creating excitement in the classroom, by C. C. Bonwell and J. A. Eison.

Coping With Difficult People, by Robert M. Bramson, PhD.

Designing Performance Appraisal Systems, Aligning appraisals and organizational realities, by Allan M. Mohrman, Jr., Susan M. Resnick-West, and Edward E. Lawler III.

Developing Critical Thinkers, Challenging adults to explore alternative ways of thinking and acting, by S. F. Brookfield.

Effective Phrases For Performance Appraisals, A guide to successful evaluations, by James E. Neal Jr.

Emotional Intelligence, by Daniel Goleman.

The Employee Strikes Back!, What you must know about sexual harassment, age discrimination, drug testing, polygraph abuse, unfair performance appraisals, wrongful firing, and more, by John D. Rapoport and Brian L.P. Zevnik.

Enhancing Adult Motivation to Learn: A Guide to Improving Instruction and Increasing Learner Achievement, by R. L. Wlodkowski.

Executive Leadership, by Jacques Clement.

The Fifth Discipline, The art and practice of the learning organization, by Peter M. Senge.

Influence, The Psychology of Persuasion, by Robert Cialdini,

Leader Effectiveness Training, The no-lose way to release the productive potential of people, by Dr. Thomas Gordon.

Leadership and the New Science, Learning about organization from an orderly universe, by Margaret J. Wheatley.

Leadership by Encouragement, by Don Dinkmeyer, Ph.D., & Daniel Eckstein, Ph.D.

In the absence of a vision, there can be no clear and consistent focus. In the absence of a dream, there can be no renewal of hope. In the absence of a philosophy, there is no real meaning to work and to life itself.

Joe Batten

Leadership For The Emerging Age, Transforming practice in adult and continuing education, by Jerold W. Apps.

Lincoln On Leadership, Executive strategies for tough times, by Donald T. Phillips.

Motivation and Personality, by A. H. Maslow.

No Fault, No Blame, How to reverse the destruction of self esteem, by C. R. Beamish.

Peak Learning, How to create your own lifelong education program, by R. Gross.

Seeing Systems, Unlocking the mysteries of organizational life, by Barry Oshry.

Self Directed Learning: A Guide for Learners and Teachers, by M. S. Knowles.

The Skills of Encouragement, Bringing out the best in yourself and others, by Dr. Don Dinkmeyer and Dr. Lewis Losoncy.

The Tao of Leadership, Lao Tzu's Tao Te Ching, Adapted for a New Age, by John Heider.

Team Building, A structured learning approach, by Peter Mears and Frank Voehl.

Training through Dialogue, Promoting effective learning and change with adults, by Jane Vella.

A Whack on the Side of the Head, How you can be more creative, by Roger von Oech.

What Followers Expect From Leaders, How to meet people's expectations and build credibility, by Kouzes & Posner. Audio Book from Jossey Bass.

To be is to do

Immanuel Kant

To do is to be.

Sigmund Freud

Dooby dooby, do.

Frank Sinatra

THE END

(Until Next Time!)

Does being creative mean you have to have NO FEAR? I can only speak for myself. I have fear—I just drag it along with me like some beloved dirty rag doll, worn out, scuffed, but familiar and needed.

Pivetta

Put Me On Your Mailing List

Professional Pride provides training books, audio skills tapes, video training, training workshops, consulting & FREE Newsletter!

☐ Catalogue ☐ Workshops ☐ Newsletter

Name ______________________________

Address ____________________________

____

____

Agency _____________________________

Address:

FAX this form to: 206-863-3568 Mail to: 1812 Pease Av, Sumner, WA 98930